ELECTRONICS AND TELECOMMUNICATIONS RESEARCH

HYDROELECTRICITY AND POWER ELECTRONICS

ENVIRONMENTAL IMPACTS, EMERGING TECHNOLOGIES AND CHALLENGES

ELECTRONICS AND TELECOMMUNICATIONS RESEARCH

Additional books in this series can be found on Nova's website
under the Series tab.

Additional e-books in this series can be found on Nova's website
under the e-book tab.

ELECTRONICS AND TELECOMMUNICATIONS RESEARCH

HYDROELECTRICITY AND POWER ELECTRONICS

ENVIRONMENTAL IMPACTS, EMERGING TECHNOLOGIES AND CHALLENGES

LUCILLE BURKE
EDITOR

nova publishers
New York

For permission to use material from this book please contact us:
Telephone 631-231-7269; Fax 631-231-8175
Web Site: http://www.novapublishers.com

Library of Congress Cataloging-in-Publication Data

Hydroelectricity and power electronics : environmental impacts, emerging technologies, and challenges / editor, Lucille Burke.
pages cm -- (Electronics and telecommunications research)
Includes bibliographical references and index.
ISBN 978-1-63321-728-7 (hardcover)
1. Hydroelectric power plants. I. Burke, Lucille.
TK1081.H895 2014
621.31'2134--dc23

2014030329

Published by Nova Science Publishers, Inc. † New York

CONTENTS

PREFACE

In modern society, the utility power grid is supposed to guarantee load management, demand side management, as well as to use the market price of electricity and forecasting of energy (e.g., based on wind and solar renewable sources) in order to optimize the whole electrical distribution system. An optimized power system is expected to have the following characteristics: high efficiency, high availability, good quality of service and high level of security, which leads to the concepts of distributed generation and smart-grid. This book discusses different emerging technologies, principles and applications of power electronics and hydroelectric power. Some of the topics discussed include parallel three-phase back-to-back converters; risk management of hydropower projects; and cascade converters for wide conversion ratios.

Chapter 1 - In Serbia, hydropower potential is considered the most important renewable resource and it is estimated at 27.2 TWh per year; technically usable power reaches 19.8 TWh per year, 18 TWh per year of which can be produced by hydropower plants larger then 10MW. Nowadays, 15 large hydropower plants produce 11.7 TWh per year. Hydropower potential is unevenly spatially distributed, which means that it is largely concentrated in only a few river basins with the potential of more than 1.000 GWh per year (the Danube, Drina, Velika Morava, Lim and Ibar rivers). On the other hand, the hydropower potential of several rivers in Serbia can only partially be exploited because their water will be used primarily as the source for regional water supply systems. The government supports growing tendencies in hydropower utilization, pointing out three priorities: 1) to modernize and upgrade the existing hydropower plants; 2) to construct new facilities and 3) encourage the development of the small hydropower sector.

The construction of hydropower plants on large rivers requires significant investment, which was an unacceptable option over the past two decades in Serbia, due to economic, political and social difficulties. On the other hand, most mountains in Serbia are relatively rich in water, with rivers running steeply downwards in short sections of the stream, which ensures favorable conditions for their exploitation for energy production and construction of small hydropower plants. Although this option is highly evaluated, numerous developing problems inhibit their construction and utilization. Accordingly, in this chapter, we present the historical development, state of the art and perspectives of the hydropower sector in Serbia. Special attention is placed on the development of hydropower plants and their impact on the environment, as well as on the cooperation between the hydropower sector and other users of water reservoirs.

Chapter 2 - Hydropower is the major renewable electricity generation technology in Europe today. It delivers storage capacity and stabilizing services for the European power system, which are crucial for a high security of electricity supply. Undoubtedly, hydropower is a mature and cost-competitive renewable energy source and a key component of the energy mix in several EU countries, and this situation will not change during the coming decades. As a multifunctional technology, it is indispensable to the electricity system, and will be even more important tomorrow, when other fossil fuels used today for the generation of electricity, such as oil and coal, will not be sufficient to satisfy the foreseeable EU energy demand.

Hydropower is expected to continue to play an important role in the world energy balance in the future, particularly in the European region. The highest growth rates, however, are expected in developing or strongly industrializing countries with high and yet unexploited hydropower potential, particularly in Eastern Europe. It is important to highlight that since the 1970s, annual energy production in some of the hydropower plants operating in Western Europe has decreased, in particular in Portugal, Spain and other Southern European countries. This reduction has been attributed to changes in average discharge, but whether this is due to temporary fluctuations or already the consequences of long-term changing climate conditions is not yet known.

Chapter 3 - There is a rapidly increasing demand for hydropower in China. However, little research has been conducted to systematically investigate the overall aspects of hydropower development risks. With support of the data collected from a fieldwork survey, this study reports the multiple facets of hydropower development risks in Southwest China as perceived by main project participants. All groups have a common view on the criticality of safety, and the groups also have their own priorities, i.e., resettlement of migrants, incompetence of subcontractors, project delay, inadequate or incorrect design, premature failure of facilities, and ecological and environmental impacts are the key risks to clients, whereas quality of work, financial related risks, reputation, and claims and disputes are the main concerns to contractors. A case study of Three Gorges Project further demonstrates that, hydropower development risks can be effectively managed by encouraging joint efforts of all participants to achieve the project goals, and hydropower in China, which is competitive in energy systems, can play a key role on producing renewable energy, reducing emissions of CO_2, and providing important social/economical benefits. Future hydropower development should emphasize the interactions between project delivery, environmental and social processes to reach appropriate trade-offs among involving stakeholders, by adequately considering the inter-relations between project participants' risks as well as hydropower project's externalities on a broad view.

Chapter 4 - Nowadays, power electronics applications in microgrids have increased their complexity. This rising of complexity implies new strategies for real time simulation and control systems design. Therefore, high performance devices are required to implement Hardware-in-the-Loop models of microgrids for real time simulations. Furthermore, control system algorithms for power electronics in microgrids demand specialized hardware with optimized architecture to execute these algorithms. Taking into account the hardware requirements, FPGAs are the best hardware option to implement Hardware-in-the-Loop models and complex control systems. In order to accomplish the mentioned requirements for power electronics in microgrids, this chapter presents in detail a method to design and implement in FPGA Hardware-in-the-Loop real time simulations and control systems in microgrids. These models and control systems are described by Systems Modeling Language (SysML). Additionally, the HiLeS-RCP (High Level Specification of embedded Systems- Rich Client Platform) methodology is employed to validate and transform the SysML diagrams in Petri nets for model implementations in FPGA. Finally, results of these implementations in embedded hardware are compared with conventional methods of simulation. To summarize, in this chapter is presented a methodology to model Hardware-in-the-Loop applications and control systems with SysML diagrams for power converters in microgrids; in addition, the methodology proposed transforms these SysML diagrams to Petri nets for their validation and implementation in FPGA.

Chapter 5 - A great number of new technological applications require DC–DC converters with wide range of conversion ratios; for example, processing energy from sustainable DC energy sources. These energy sources, before being connected to the grid, require converters with a significant step-up conversion ratio. On the other hand, there are technological developments that require converters with high-input voltage and low-voltage/high-current output specifications. For example, the new generation of microprocessors is being designed to operate using power supplies with an output voltage from 3.3 V to 1 V or less. A first approach to provide a significant voltage conversion is to operate a switching converter at a very high or small duty cycle. However, the maximum and minimum attainable conversion ratios for the conventional converters are limited by the characteristics of the switching devices. An often-used approach is to use step-down or step-up transformers; however, large switching surges due to leakage inductance may damage the switching devices and make the controller difficult to design. A transformerless topology that provides a wide conversion ratio is the cascade converter, which consists in two or more basic switch-mode DC–DC converters connected in cascade where the voltage ratio is a function of the number of stages. In this chapter, a discussion of the operation of cascade converters with a single active switch is given. The operating conditions are developed together with the corresponding models. The resulting models are compatible with the standard techniques widely used in control design.

Chapter 6 - High performance normally-off GaN-based metal oxide semiconductor field effect transistor (MOSFET) was fabricated. The buffer layer of the MOSFET was grown by varying the growth temperature to control the size of nucleation sites which results in extremely high buffer resistance($>10^{12}$ Ω/sq). The fabricated small-area MOSFET exhibited excellent normally-off device characteristics, such as threshold voltage of 2 V, maximum drain current of 253 mA/mm, on-off current ratio of 5.5×10^7, destructive breakdown voltage of 830 V, and leakage current of 0.7 μA/mm at V_{DS} of 600 V. The corresponding values of the large-area MOSFET with multi-finger pattern were 0.6 V, 6 A, 1.3×10^7, 670 V, and 50 μA (0.28 μA/mm).

Chapter 7 - This chapter proposes bidirectional three-phase to three-phase backto-back converters, which includes their generalization with *n* units in parallel. Such converters can be employed to interface high power renewable energy sources (e.g., wind turbines) to the utility grid. Although the proposed converters are characterized by parallel connection without use of transformers, their control system permits canceling the circulating current among the units in parallel. In addition to proposing new configurations, this chapter presents a comprehensive model of the topologies, and a suitable control strategy for the system with regulation at both sides of the converter. Compared to the conventional three-phase to three-phase back-to-back converter, the proposed topologies allow the following advantages: i) reduction on the current and consequently power ratings of the power switches, ii) reduced harmonic distortion at both sides of converter, and iii) lower dc-link capacitor losses, which has a positive impact on the lifespan of the power system. Additionally, the semiconducto r losses can be reduced when the proposed converter is used. Simulated and experimental results are also presented for validation of the theoretical expectations.

In: Hydroelectricity and Power Electronics
Editor: Lucille Burke
ISBN: 978-1-63321-728-7

Chapter 1

HYDROPOWER DEVELOPMENT IN SERBIA: CURRENT STATE, PERSPECTIVES AND CHALLENGES

Milena Panić, Marko Urošev, Ana Milanović Pešić and Dragana Miljanović*

Geographical Institute "Jovan Cvijić" of the Serbian Academy of Sciences and Arts, Belgrade, Serbia

ABSTRACT

In Serbia, hydropower potential is considered the most important renewable resource and it is estimated at 27.2 TWh per year; technically usable power reaches 19.8 TWh per year, 18 TWh per year of which can be produced by hydropower plants larger then 10MW. Nowadays, 15 large hydropower plants produce 11.7 TWh per year. Hydropower potential is unevenly spatially distributed, which means that it is largely concentrated in only a few river basins with the potential of more than 1.000 GWh per year (the Danube, Drina, Velika Morava, Lim and Ibar rivers). On the other hand, the hydropower potential of several rivers in Serbia can only partially be exploited because their water will be used primarily as the source for regional water supply systems. The government supports growing tendencies in hydropower utilization, pointing out three priorities: 1) to modernize and upgrade the existing

* Corresponding author: m.panic@gi.sanu.ac.rs.

hydropower plants; 2) to construct new facilities and 3) encourage the development of the small hydropower sector.

The construction of hydropower plants on large rivers requires significant investment, which was an unacceptable option over the past two decades in Serbia, due to economic, political and social difficulties. On the other hand, most mountains in Serbia are relatively rich in water, with rivers running steeply downwards in short sections of the stream, which ensures favorable conditions for their exploitation for energy production and construction of small hydropower plants. Although this option is highly evaluated, numerous developing problems inhibit their construction and utilization. Accordingly, in this chapter, we present the historical development, state of the art and perspectives of the hydropower sector in Serbia. Special attention is placed on the development of hydropower plants and their impact on the environment, as well as on the cooperation between the hydropower sector and other users of water reservoirs.

Keywords: hydropower potential, hydropower plants, Serbia

INTRODUCTION

Every nation strives to fulfill its needs for energy by using primarily natural resources that are more or less nationally available. As these needs have grown significantly over the past decades due to technical, technological, economic and social development, the awareness of the need to protect environmental quality and ensure a rational utilization of resources has become globally present. While discussing this issue, Espinoza and Barragán (2013) point out that the energy sector is an agent of economic development but can also be a source of social and environmental conflict.

The attitude that fossil energy resources should be depleted and that the share of alternative, environmentally friendly and justified, energy sources in the global energy balance should rise has gained importance. However, most renewable energy technologies continue to play a modest role in the global energy matrix due to their high production costs, compared to conventional technologies which use fossil fuels. This is a consequence of neglecting environmental externalities in energy cost considerations [1].

Hydropower is one of such energy sources; it is renewable and ecologically pure. Having in mind the capacity for multipurpose use, utilizing water power is usually the most rational way of producing electric power. Hydropower is statistically the most common source of renewable energy and it plays a very important role in the global energy production; however, it strongly depends on a country's geography [2]. According to the International Energy Agency (IEA), the global generation of electric power from hydropower in 2010 was 3402.3 TWh, which accounted for almost 17% of the total global energy production. About 70% of economically feasible hydropower potential worldwide remains undeveloped [3]. As Zimny et al. (2013) have demonstrated in their study on the development of hydropower in the world between 1995 and 2011, the global technical potential is estimated at more than 16,400 TWh/year. For Europe, the technical potential is estimated at 1,021 TWh/year. Also, the utilization of hydropower potential for electric power generation in Europe shows a continuous rising trend, as a result of new investments in the Nordic countries. European countries with the largest production in the mentioned period were Norway, Sweden and France. Also, the installed capacities of SHPs in the EU and the world have increased from 5,900 MW (EU) and 19,000 MW (world) in 1980 to 10,300 MW (EU) and 46,000 MW (world) in 2005 [4]. The amount of electrical energy produced in SHPs in EU was expected to reach 55 TWh in 2010, whereas their combined capacity was expected to be 14 GW. According to data provided by the European Small Hydropower Association (ESHA), Austria has the greatest share of SHPs in the total installed capacity - 4.89%, whereas Germany has the greatest share of SHPs in the total capacity of hydropower stations–16.67%. Among the countries that have recently joined the EU, the greatest number of SHPs have been constructed in the Czech Republic, Poland and Slovenia [5]. Scientists of various disciplines (ecologists, hydroengineers, economists, etc.) debate over the advantages and disadvantages (strengths and weaknesses) of hydropower projects, i.e., the utilization of watercourses for hydropower production; this involves the construction of dams and impoundment of water.

Today, the power system of Serbia has a nominal capacity of about 9,100 MW at plant gate; the share of hydropower plants is about 30%, while the rest is the contribution of thermal power plants. The point of departure highlighted in adopted strategic and planning documents is the conclusion that energy resources are not sufficiently explored, while available data are not conclusive; however, there is a general agreement that besides geological reserves of primary energy resources, hydropower potential and other renewable energy resources will be the basis for the development of production capacities in the future. In 2009, 42% of the total primary energy requirements of Serbia were covered by import with a growing trend of import dependency [6]. However, in order to ensure a sustainable energy development in Serbia, it is necessary to increase the utilization efficiency of locally available non-renewable and renewable resources [7].

Since the early 20th century, when electric power from hydropower was used mainly for lighting and the needs of local industry, Serbia has had a long tradition of hydropower potential exploitation. The earliest hydropower plants in Serbia had a smaller capacity and were built close to consumers, due to their limited long-distance power transmission capacity. The first hydropower plant in Serbia was built on the Rasina River (the Morava River basin) in the town of Kruševac, in 1891. The first hydropower plant based on Nikola Tesla's principle of three-phase alternating current was the plant Pod Gradom on the Đetinja River (the Morava River basin) in the town of Užice. It was installed in 1900, only four years after the first such hydropower plant at Niagara Falls in the United States. Several dozen hydropower plants of this type had been installed before the focus was switched to the construction of plants on large rivers. The lines of development followed after World War II implied an intensive hydropower development and construction of large hydropower plants. Further study of the hydropower potential of small watercourses was abandoned–unjustifiably, having in mind their potential. That period had lasted until 1990, when the last large hydropower plant, the HPP Pirot on the Visočka River, was built in Serbia. Due to unfavorable economic, political and social factors, significant investment in new major hydroelectric facilities was impossible. Since then, the idea of small hydropower plants has once again become topical as a good approach to fostering hydropower utilization; it has been highlighted in all strategic documents and has been supported by the government.

Depending on the mode of operation and the availability of additional facilities necessary for an effective utilization of available resources, hydropower plants in Serbia can be divided into *storage hydropower plants* (where the natural flow regime of a river may be, to a certain extent, adjusted to the needs of the power system); *run-of-river hydropower plants* (in which the natural flow regime is used for electric power production) and *reversible or pumped storage hydropower plants* (where water is artificially pumped into a reservoir which is at a much higher elevation than the river).

In Serbia 64% of hydropower plants are run-of-river plants, 15% are storage plants, whereas 21% are reversible (pumped storage) plants. The power of thermal plants and run-of-river hydropower plants accounts for almost 90% of the total capacity, while their share in electric power production reaches 95%. These facts, as well as an extremely uneven consumption, lead to great difficulties in fulfilling consumers' needs for electric power during periods of peak consumption [8].

In recent years, the following classification of small hydropower plants based on their combined power capacity and the capacity of a single unit is becoming widely adopted:

- up to 0.01 MW–pico HPs;
- up to 0.1 MW–micro HPs;
- up to 1 MW–mini HPs; and
- up to the combined power capacity reaching 30 MW and the capacity of a single power unit reaching 5 MW–small HPs.

In the last case, the maximum capacity of a single unit is sometimes limited to 10 MW rather than 5 MW [5]. According to Serbian regulations, until December 2012, the term small hydropower plants covered all hydropower plants with the installed power up to 10 MW regardless of their type (i.e., it included both plants using reservoirs and run-of-river hydropower plants). However, since the adoption of the Decree on Incentive Measures for Privileged Electric Power Producers [9] in January 2013, this term has been extended to include hydropower plants with the installed power up to 30 MW. The hydropower plants with the installed power up to 100 kW are called microenergy plants [10]. Apart from power, SHPs may also be classified according to other criteria (penstock, technological concept, linkage with a network, etc.).

The Government of the Republic of Serbia supports trends aimed at increased hydropower; it has defined three priorities: the modernization and upgrading of existing hydropower plants, the construction of new facilities and encouraging the development of the small hydropower sector.

2. Legal, Institutional and Planning Framework for the Production and Exploitation of Hydropower in Serbia

Having in mind that the energy sector has a strategic significance in every country and that it has an important impact on economic and sustainable development, it is necessary to ensure its legislative, institutional and planning regulation. In Serbia, the attitude towards the energy sector has been influenced by various factors, which have, to a certain extent, caused it to take a particular form. An economic development accompanied with difficulties and major political changes have led to frequent changes of responsible institutions, laws and regulations and have made it impossible to monitor the planned developments. Also, the established legal framework has been subject to permanent improvement and harmonization with the legislation of the European Union. One of the most significant step on this path was implementation of Water Framework Directive (Directive of European Parlament and of the Council 2000/60/EC). As well as the ratification of the Treaty establishing the Energy Community, by which the Republic of Serbia agreed to adopt and carry out the plan for the implementation of the Directive 2001/77/EC on the promotion of electricity produced from renewable energy sources, to enforce a set of regulations on climate change aimed at reducing greenhouse gas emissions and to accede the International Renewable Energy Agency (IRENA) [7, 11].

The Energy Law of the Republic of Serbia (Official Gazette of the Republic of Serbia 84/04, 57/2011) is the main document which regulates "the energy policy objectives and the methods of its implementation; requirements for a reliable, safe and efficient delivery of power and energy-generating materials; conditions for the reliable supplying of customers; conditions for the construction of new energy facilities; conditions and methods of performing energy-related activities; organization and functioning of the electric power and natural gas market; rights and obligations of market participants; protection of the consumers of energy and energy-generating materials; terms, conditions and incentives for energy production from renewable energy sources and the combined production of heat and power; rights and responsibilities of government agencies; status, funding, activities and other issues relevant to the work of the Energy Agency of the Republic of Serbia related to monitoring the implementation of this law." There are two basic characteristics of the Energy Law. The first is the separation of responsibilities for the adoption of the energy policy, strategy and other regulations from the jurisdiction over their implementation. The second is the reorganization energy companies, the abolition of monopolies and the introduction of market competition in those energy activities in which competition can actually exist. In those sectors where this is not possible due to natural monopolies, the Energy Agency shall provide strict regulation and control of the monopolistic activities [12]. Furthermore, in a separate segment, the Energy Law regulates issues related to renewable energy sources, pointing out that it is in the interest of the Republic of Serbia to utilize them. Small hydropower plants in the basic and general sense are also covered in this segment. The Law regulates the issue of small hydropower plants, all types of licenses and permits and their privileged position in the market, compared to other energy producers who sell energy under equal conditions; it also implies a right to subsidies (tax, tariff and other subsidies provided for by law), as well as incentive feed-in tariffs.

Other legal documents regulate in detail particular issues, such as granting concessions, obligations regarding environment protection, methods of construction of facilities and operation of public companies. They include the Law on Planning and Construction (Official Gazette of the RS 72/09, 81/2009), Law on Environmental Protection (Official Gazette of the RS 135/04, 36/2009), Law on Waters (Official Gazette of the RS 30/2010), Law on Concessions (Official Gazette of the RS 55/2003), Law on Public Enterprises and Performing Activities of General Interest (Official Gazette of the RS 25/2000, 25/02, 107/05, 108/05) and other legal acts. There are also by-laws intended to determine conditions under which energy is supplied and ensure that the rules of operation be respected by all entities involved in the energy industry (Decree on the Requirements for Obtaining the Status of a Privileged Electric Power Producer and the Criteria for Assessing the Fulfillment of these Requirements; Model Agreement on Purchasing Electric Power from Privileged Producers, adopted in 2009; and the Decree on Incentive Measures for Privileged Electric Power Producers, adopted in 2013.)

The institutional framework includes responsible institutions whose primary task is to enable the drafting, adoption and implementation of different laws and regulations in the energy sector. The Ministry of Mining and Energy is the supreme institution which conducts various activities related to energy utilization. Its main tasks are to prepare and propose the Energy Development Strategy, as well as to monitor its implementation; determine the criteria, conditions, terms and procedures for obtaining the status of privileged energy producers; prescribe the manner and conditions for issuing energy permits and maintain a register of authorizations; draft and propose national action plans and monitor their implementation; draft and propose the public tender call for the construction of facilities for electric power production, etc.

The Energy Agency is a regulatory body established with the idea of promoting and directing the development of electric power and natural gas production in compliance with the principles of non-discrimination and effective competition by creating a stable regulatory framework; it shall also perform other duties stipulated by the Energy Law. The main tasks of the Energy Agency are to devise the methodology for determining prices for public supply of electric power, electric power transmission and distribution, connecting consumers to the electric power transmission and distribution system, and licensing for performing energy-related activities, etc. [12].

Other responsible institutions include the Ministry of Agriculture and Environmental Protection, Ministry of Economy, Ministry of Construction, Transport and Infrastructure, Provincial Secretariat for Energy and Mineral Resources of the Autonomous Province of Vojvodina, local self-government units and regional centers for energy efficiency.

The Energy Development Strategy of the Republic of Serbia until 2015 has been devised with the aim of identifying various energy sources in Serbia, with special emphasis on hydropower potential and the modes and possibilities for its utilization in future. The Strategy draws attention to the hydropower potential of Serbia–both the hydropower potential of large rivers, such as the Morava, Drina, Lim and the Danube, and the hydropower potential of small rivers, especially in rural and mountainous areas. On large rivers, the construction of individual facilities with a capacity greater than 10 MW is foreseen, while on small rivers, small hydropower plants, as the most important renewable energy source in Serbia, are to be set up.

The Spatial Plan of the Republic of Serbia from 2010 to 2020 (Official Gazette of RS 88/10) is the main document governing the spatial development of the Republic of Serbia in the mentioned ten-year period. The Plan provides an overview of the available hydropower potential of Serbia by linking it with the basic guidelines for energy development in the future. It also highlights priorities, such as the existing energy facilities on existing water reservoirs, multipurpose reservoirs within the water management network, the construction of new hydropower plants that would exploit the potential of boundary river basins and the construction of pumped storage hydroelectric power plants on major rivers in Serbia. The Plan also deals with potential locations where small hydropower plants could be built in the territory of Serbia, taking into account documents such as the *Survey of Small Hydropower Plants in the Republic of Serbia* (1987) and the *Survey of Small Hydropower Plants in Vojvodina* (1989) and it also suggests that such locations be protected against unplanned usage. Although the Survey of Small Hydropower Plants in the Republic of Serbia (1987) and the Survey of Small Hydropower Plants in Vojvodina (1989) are not officially current, they are still used as supplement to official legal documents in the process of choosing potential locations for the construction of SHPs. The *Survey of Small Hydropower Plants in the Republic of Serbia* (1987) and the *Survey of Small Hydropower Plants in Vojvodina* (1989) were drafted in the late 1980s. These documents include a detailed study of the energy potential of watercourses and selected locations for the construction of SHPs. According to the mentioned surveys, 856 locations suitable for the construction of such facilities were identified in Serbia and additional thirteen in Vojvodina. Due to turbulent economic and political changes that followed in the 1990s, the ideas underlying these studies were not put into practice. According to the results of fifteen recently completed preliminary feasibility studies on the construction of small hydropower facilities, under the present economic conditions, it would be possible to install them on 5–10% of the total locations foreseen by the Survey. These studies also include elements of a draft project.

Table 1. Natural hydropower potential [8]

River	Average discharge (m^3/s)	Average head (m)	Potential (GWh/year)	Length (km)	Specific potential (GWh/km·per year)
(1)	(2)	(3)	(4)	(5)	(6)
Dunav	5657	38	10000.00	260	38.46
Sava	1642	4	463.60	115	4.03
Kolubara	21.9	989	146.43	132	1.11
Drina*	413	186	5678.41	210	27.04
Jadar	7.92	388	42.67	47	0.91
Lim	97.5	242	1583.81	109	14.53
Uvac	17.1	1160	937.00	124	7.56
Ibar	58.5	1018	997.74	134	7.45
Raška	7.35	350	65.88	39	1.69
Studenica	5.19	732	151.79	49	3.10
Jošanica	1.82	544	47.78	21	2.27
Moravica	18.2	501	198.78	78	2.55
Veliki Rzav	7.43	582	201.81	57	3.54
Đetinja	10.8	489	48.25	71	0.68
Skrapež	5.25	138	30.23	40	0.76
Kamenica	1.47	324	25.84	40	0.65
Zapadna Morava	119	168	766.58	211	3.63
Rasina	8.35	540	134.04	93	1.44
Moravica (Preševska)	0.95	320	5.04	30	0.17
Južna Morava	97.6	354	797.32	316	2.52
Vrla	2.1	747	51.93	27	1.93
Vlasina	10.1	954	265.19	59	4.49
Nišava	27.9	299	429.68	151	2.84
Jerma	5.69	211	74.29	31	2.40

Table 1. (Continued)

River	Average discharge (m^3/s)	Average head (m)	Potential (GWh/year)	Length (km)	Specific potential (GWh/km·per year)
(1)	(2)	(3)	(4)	(5)	(6)
Temštica (Visočica)	8.21	510	199.77	72	2.78
Veternica	8.92	458	53.85	75	0.72
Jablanica	6.41	180	53.14	84	0.63
Pusta reka	2.83	344	22.71	66	0.34
Toplica	17.6	725	277.34	124	2.24
Moravica (Aleksinačka)	3.86	287	48.31	57	0.85
Velika Morava	216	64	1090.33	221	4.93
Resava	4.89	571	79.68	85	0.94
Crnica	2.02	55	16.11	44	0.37
Mlava	10.5	653	130.92	145	0.90
Pek	10.3	712	198.86	134	1.48
Timok	30.2	525	275.90	169	1.63
Crni Timok	12.4	282	151.47	75	2.02
Svrljiški Timok	4.99	490	76.64	59	1.30
Beli Drim	60.0	230	1230.00	108	11.38
Lepenac	7.00	685	134.00	60	2.23

* figures cover only hydropower potential in Serbia.

3. Natural and Utilized Hydropower Potential of Serbia

In Serbia, hydropower potential is considered the most important renewable resource and it is estimated at 27.2 TWh per year; technically usable power reaches 19.2 TWh per year, 17.5 TWh per year of which can be produced by hydropower plants larger than 10 MW. Nowadays, 15 large hydropower plants produce 11.7 TWh per year.

Table 2. Existing hydropower plants [8, modified]

No	Hydropower plant	River	Year of construction	Q_{sr} (m^3/s)	Q_i (m^3/s)	H_i (m)	P_i (MW)	W (GWh)	HPP type
(1)	(2)	(3)	(4)	(5)	(6)	(7)	(8)	(9)	(10)
1	Đerdap I	Dunav	1970/72	5668	4800	27	1026	5650	r*
2	Đerdap II	Dunav	1985/87	5668	4200	8	270	1332	r*
3	Zvornik	Drina	1955	389	620	20	96	515	r*
4	Bajina Bašta	Drina	1966/68	354	600	65	365	1691	r*
5	RHE Bajina Bašta	Drina+ B. Rzav	1982	/	129/ 104	572/579	614/ 616	1200/ 1600	r***
6	Uvac	Uvac	1979	10.2	43.0	100	36	72	s**
7	Kokin Brod	Uvac	1962	13.6	37.4	54	22	60	s**
8	Bistrica	Uvac	1960	14.4	36.0	360	104	370	s**
9	Potpeć	Lim	1967/70	91.9	165	38	52	201	r*
10	Gazivode	Ibar	1981	14.5	42.6	87	35	70	s**
11 12 13 14	Vlasinske elektrane (Vrla I-IV)	Vlasina + Vrla + Božička reka	1955/75	6.50	18.3 18.5 18.5 18.5	338 152 185 164	51 24 29 25	347	s**
15	Pirot	Visočica	1990	3.08	45.0	206	80	122	s**

*run-of-river plants; **storage plants; *** reversible plants;

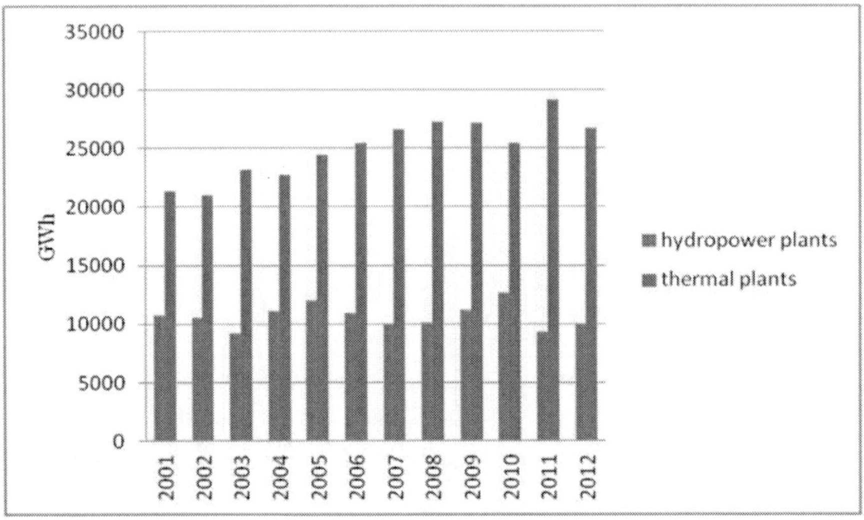

Figure 1. Electric power production in Serbia in the period 2001-2012 (Source of data: [13]).

Hydropower potential is unevenly spatially distributed, which means that it is largely concentrated in only a few river basins with the potential of more than 1.000 GWh per year (the Danube, Drina, Velika Morava, Lim and Ibar rivers). The hydropower potential of the forty largest rivers in Serbia is presented in Table 1.

Out of the total exploitable hydropower potential by facilities larger than 10 MW, about 60% has so far been utilized, taking into account a half of the potential in watercourses shared with the former Yugoslav republics [8]. An overview of existing hydropower plants (listed according to river basins) involved in the power transmission network and their basic parameters are presented in Table 2. Table 2 shows that attention has so far been mostly focused on exploiting rivers with the largest hydropower potential (the Danube and Drina rivers) and some smaller watercourses favourable for hydropower use: the Vlasina, Lim, Visočica and other rivers. The annual electric power production by hydropower plants varies depending on the hydrological situation and its share in the total electric power production in Serbia reaches about 30% (Figure 1). Besides the construction of large hydropower plants over 10MW, hydropower utilization in Serbia involves the construction of small hydropower plants. Most mountains in Serbia are relatively rich in

Table 3. Small hydropower plants in Serbia

No	Name	River	Municipality	Owner	Power (MW)
1	Radaljska banja	Radalj	Mali Zvornik	EPS*	0.27
2	Đorđić	Ljubovića	Ljubovija	Private	0.01
3	Vrelo	Perućačko vrelo	Bajina Bašta	EPS	0.06
4	Vrutci	Đetinja	Užice	Private	0.40
5	Turica	Đetinja	Užice	EPS	0.32
6	Pod gradom	Đetinja	Užice	EPS	0.36
7	Ovčar banja	Zapadna Morava	Čačak	PPC	6.00
8	Međuvršje	Zapadna Morava	Čačak	PPC	7.00
9	Kratovska reka	Kratovska reka	Priboj	EPS	1.50
10	Seljašnica	Seljašnica	Prijepolje	EPS	1.26
11	Moravica	Golijska Moravica	Ivanjica	EPS	0.16
12	Studenica	Studenica	Kraljevo	SOC**	0.09
13	Radošićska reka	Radošićska reka	Raška	Private	0.04
14	Raška	Raška	Novi Pazar	EPS	0.80
15	Kuršumlija	Toplica	Kuršumlija	Private	0.35
16	Grčki mlin	Toplica	Prokuplje	Private	0.08
17	Sokolovica	Veliki Timok	Zaječar	EPS	5.20
18	Gamzigrad	Crni Timok	Zaječar	EPS	0.22
19	Jevtić	Crni Timok	Zaječar	Private	0.10
20	Bovan	Sokobanjska Moravica	Aleksinac	Private	0.26
21	Elektro Slavica	Trgoviški Timok	Knjaževac	Private	0.05
22	Sićevo	Nišava	Niš	EPS	1.35
23	Sveta petka	Nišava	Niš	EPS	0.60
24	Temac	Temštica	Pirot	EPS	0.78
25	Tegošnica	Vlasina	Crna Trava	Private	0.64
26	Poštica	Poštica	Vlasotince	Private	0.70
27	Livađe	Darkovačka reka	Crna Trava	Private	0.45
28	Vučje	Vučjanka	Leskovac	EPS	0.93
29	Munja	Vrla	Vladičin Han	Private	0.03
30	Jelašnica	Jelašnica	Vladičin Han	EPS	0.40
31	Prvonek	Banjska	Vranje	EPS	0.90
32	Krstići	Vlasina	Crna Trava	Private	0.70
33	Manastir Rača	Rača	Bajina Bašta	SOC	0.06
34	Prisoje	Tripušnica	Trgovište	Private	0.91
35	Crkvine	Bistrica	Nova Varoš	Private	0.96

* EPS–Electric Power Industry of Serbia

** SOC–Serbian Orthodox Church

water, with rivers running steeply downwards in short sections of the stream, which ensures favourable conditions for their exploitation for energy production. Although all preconditions in terms of the position of rivers, topographic and hydrological conditions are met, the attention paid to this problem is still insufficient and there is a lack of action in dealing with it [14]. The potential of small watercourses suitable for installing small hydropower plants reaches 0.4 million tonnes of oil equivalent or 3% of the total potential of renewable sources in Serbia [8]. So far, 73 SHPs have been built on the rivers in Serbia, but the majority of them are in poor condition. Out of that number, 35 with a total capacity of 33.60 MW and an annual electricity production of approximately 180 GWh are in use (Table 3). 38 facilities with a total power of 8.667 MW are out of use [15]. The Electric Power Industry of Serbia is in charge of 18 SHPs and they are included in the national electrical energy system, whereas the remaining 17, which are independent, have the status of privileged electricity producers.

3.1. Overview of the Hydropower Plants Larger Than 10 MW in Serbia

The existing hydropower plants (HPPs) are located in the basins of the Drina and Velika Morava rivers, as well as on the Danube River (Figure 2). In this section, we will present a detailed overview of reservoirs and hydropower plants in the mentioned basins which are larger than 10MW.

The Danube River

The Danube River is 2857 km long and it is the second largest river in Europe. In Serbia, it is 588 km long (20.6% of total length) and it has a great hydropower potential that is not sufficiently used. The Danube River has many tributaries in Serbia; mean multi-annual discharge is approximately 5650 m^3/s [16].

The hydropower and navigation system Đerdap I (Iron Gate I; Figure 3) is the largest hydropower plant in Serbia and the largest hydro-engineering facility on the Danube River. The HPP Đerdap I is located at a distance of 943 km from the mouth of the Danube, 10 km upstream from Kladovo. In mid-1964, preparation works for its construction were begun; in 1969, the first lock was opened for traffic, while in 1970, the first hydro-generating units were put into operation; in 1971, the Đerdap I system was ready for full exploitation. On the Serbian side, six hydropower generating units were set up. The individual capacity of each unit is 171 MW (the total power of all six units was 1026 MW). They have an installed flow rate of 4800 m^3/s, a calculated net fall of 27.16 m and an average annual production of 5489 GWh [17]. The total volume of the Đerdap I reservoir is 2800×10^6 m^3, the maximum backwater elevation is 69.50 m above sea level, the average reservoir inflow is 5650 m^3/s, whereas the size of the Danube catchment area up to the reservoir is 577×10^3 km^2. The two-level ship lock is 310 m long and 34 m high; its depth at the threshold is 5.5 m; it has an under-keel clearance of 13.5 m and is suitable for river and sea ships with a carrying capacity of 5000 t. The total length of the structure is 1278 m, whereas its height is 34.8 m. It is fully symmetrical and designed so that Serbia and Romania have identical parts of the main structure and they maintain and use them in compliance with agreements and conventions related to the construction and exploitation of the power plant. The bisector of the structure coincides with the state border. Each country owns one power plant, one ship locks and seven spillways [17].The two power plants are connected so that power generators on the Serbian side can deliver electric power to the network on the Romanian side, and vice versa, if such a need arises.

The Đerdap II (Iron Gate II) hydropower plant is located 80 km downstream from the HPP Đerdap I, at a distance of 863 km from the mouth of the Danube, at the profile Kusjak–Ostrovul Mare. This is another power plant on the Danube River that was jointly constructed by Serbia and Romania. This system, completed in 1984, is a multi-purpose hydro-engineering facility, just like the HPP Đerdap I. It consists of a main power plant, two additional power plants, two ship locks, two overflow dams and two power distribution facilities. These facilities are equally shared between Romania and Serbia. The total installed capacity of the power generating unit on the Serbian side is 270 MW (10×27 MW), whereas the installed flow rate is 4200 m^3/s. The turbines were successively put into operation between 1985 and 2001. The total volume of the reservoir is 716.5×10^6 m^3 [17].The Serbian part of the Đerdap I and Đerdap power plants are used by the Electric Power Industry of Serbia, namely the HPP Đerdap Joint-Stock Company (HE Đerdap d.o.o.). According to data provided by the Electric Power Industry of Serbia, the average annual production is around 1500 GWh.

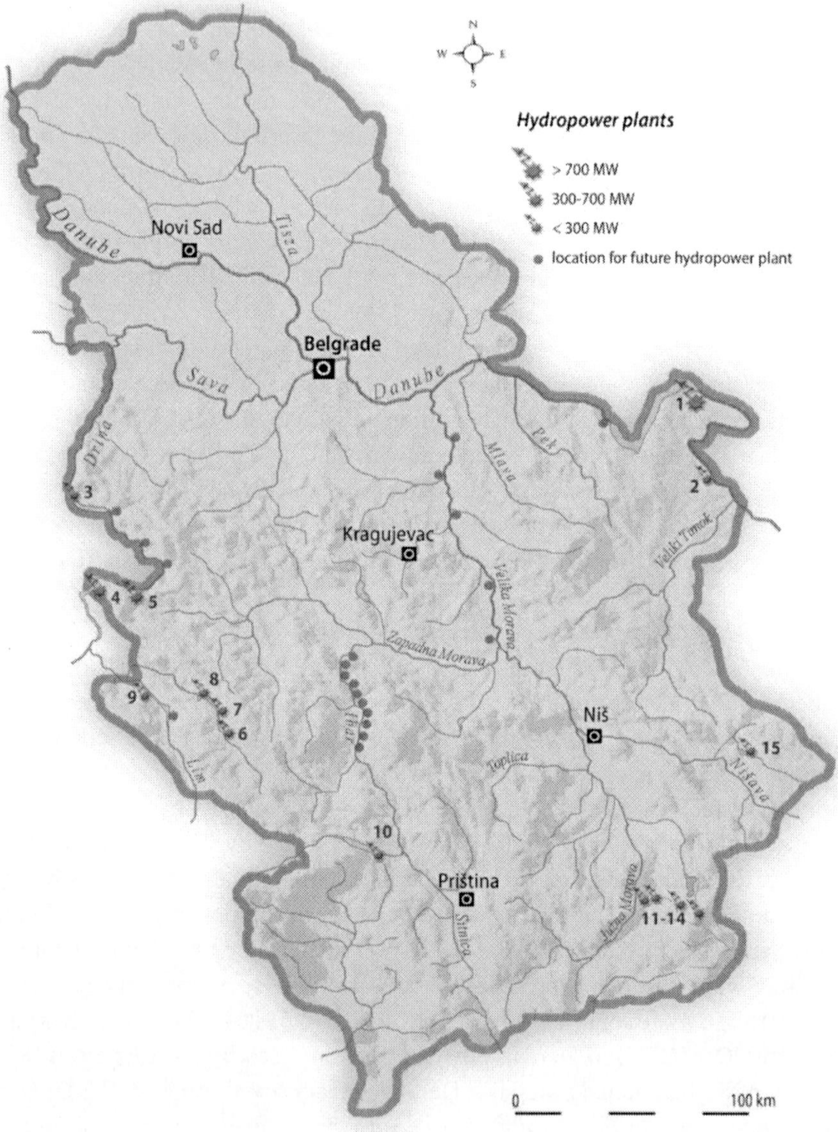

Figure 2. Existing hydropower plants and locations for new hydropower plants in Serbia (also see Table 2).

Figure 3. Đerdap I hydropower plant on the Danube River (photo by Ž. Nikolić, 2003).

The Drina River Basin

The Drina River is the one of the most important underutilized hydropotential in the region. The entire Drina River basin with an area of 19,570 km^2 and is shared between three neighboring states: Serbia, Bosnia and Herzegovina and Montenegro. Out of the whole basin area, 30.5% belong to Serbia. The average rainfall is about 1,100 mm with extremely irregular water regime [18]. The water resources in the Drina River basin are mainly used for hydropower production.

In the Drina River basin, there are nine hydropower plants: seven in Serbia (HPP Zvornik, HPP Bajina Bašta, Reversible HPP Bajina Bašta, HPP Bistrica, HPP Kokin Brod, HPP and HPP Uvac Potpeć), one in Bosnia and Herzegovina (HPP Višegrad) and one in Montenegro (HPP Piva). The total installed capacity of these facilities is 1980 MW, whereas their average annual production is 6320 GWh [19]. All of the hydropower plants in the territory of Serbia are used by the Electric Power Industry of Serbia, namely by the Drina-Lim Hydropower Plants Joint-Stock Company (Drinsko-Limske Hidroelektrane d.o.o.).

The Zvornik concrete gravity dam is located in the lower course of the Drina River (at a distance of 93 km from the river mouth). It was built between 1948 and 1955. The catchment area of the Zvornik reservoir is 17,423 km^2; the average flow rate is 389 m^3/s, while the total volume of the reservoir is 89×10^6

m^3. The Zvornik hydropower plant is a run-of-river facility with two independent powerhouses: one on the left and the other on the right bank of the Drina River. Between the powerhouses, there is a concrete gravity dam, 167 m long, with eighth spillways and a capacity of 7 800 m^3/s [19]. The HPP Zvornik, with a capacity of 96 MW (4×24 MW), an installed flow rate of 620 m^3/s and a calculated net fall of 20.3 m, has an average annual production of 515 GWh [20]. The complex of facilities includes a fish path and a log canal, which significantly mitigates the environmental consequences of the dam on the river ecosystem.

The Bajina Bašta dam (Figure 4) was built between 1959 and 1966 in the middle course of the Drina River (at a distance of 199 km from the river mouth), 15 km upstream from the town of Bajina Bašta, with the primary purpose of producing electric power. This is a lightweight gravity dam with cellular buttresses; it is 461 m long, 90 m high, and has five spillways with a capacity of 11,600 m^3/s. The catchment area of the Bajina Bašta reservoir is 15,195 km^2; it has an average flow rate of 349 m^3/s, and a total volume of 340×10^6 m^3. The reservoir is 50 km long and has a useful volume of 218×10^6 m^3, enabling weekly and, partially seasonal water-level adjustment. The Bajina Bašta hydropower plant is a facility with an independent powerhouse; it has a power of 364 MW (4×91 MW), an installed flow rate of 600 m^3/s and a calculated net fall of 65 m [19]. According to the Water Master Plan of Serbia, it has an average annual production of about 1691 GWh.

Between 1976 and 1982, on mountain Tara, in the basin of the Beli Rzav River, uphill from the village of Lazići, an eponymous dam was built and a reservoir known as Zaovine Lake was formed. It has a total volume of 170×10^6 m^3 and is used as a compensation reservoir by the pumped storage (reversible) hydropower plant Bajina Bašta [21]. The capacity of the reservoir ensures water-level adjustment during several months. This is an earth-fill embankment dam with a clay core; it is 218 m long and 130 m high, with a spillway capacity of 170 m^3/s. The derivation unit is housed in a separate powerhouse near the Bajina Bašta hydropower plant (which has an independent powerhouse), while the reservoir of the HPP Bajina Bašta is used as a lower reservoir. The water transport system consists of a supply tunnel and pipelines with a surge tank. According to a technical report of the Drina-Lim Hydropower Plants company, the reversible hydropower facility is powered by two turbines with a total power of 614 MW (616 MW in pump mode), an installed flow rate of 129 (104) m^3/s, a calculated net fall of 572 (579) m and an average annual production of 1200 (1600) GWh [20].

In the Lim River basin, which belongs to the Drina River basin and has a large hydropower potential, there are four reservoirs for electric power production: three on the Uvac and one on the Lim River.

Figure 4. Bajina Bašta hydropower plant on the Drina River (photo by M. Urošev, 2013).

The Radoinja dam was built in 1959 on the Uvac River (at a distance of 62 km from the river mouth), 12 km downstream from the Kokin Brod dam. This is a rock-fill embankment dam with screen from cast asphalt; it is 361 long and 42 m high and has three spillways with a capacity of 1400 m^3/s. The catchment area of the Radoinja reservoir is 1500 km^2; the average flow rate is 14.4 m^3/s, while the reservoir's total volume is 7.6×10^6 m^3 [19]. Through a tunnel, water is taken from the Radoinja reservoir to a surge tank and, further, via two pipelines under pressure, to the HPP Bistrica, a derivation unit on the right bank of the Lim River. The HPP Bistrica is a storage power plant with a capacity of 104 MW (2×52 MW), an installed flow rate of 36 m^3/s, a calculated net fall of 360 m and an average annual production of 370 GWh [19].

The Kokin Brod dam was built in 1962 on the Uvac River (at a distance of 54 km from the river mouth), at the 11th kilometre of the road connecting Nova Varoš and Zlatibor mountain. This is an earth-fill embankment dam with a clay core. A spillway with a capacity of 1400 m^3/s divides the dam into the main dam (500 m long and 82 m high, located in the river bed) and a side dam (720 m long and 25 m high), which bars the saddle on the left river bank. The catchment area of the Kokin Brod reservoir is 1170 km^2; the average flow rate is 13.9 m^3/s, while the total volume is 250×10^6 m^3 [19]. The Kokin Brod hydropower plant has an independent powerhouse, a capacity of 22.5 MW (2×11.25 MW), an installed flow rate of 37.4 m^3/s, a calculated net fall of 54 m and an average annual production of 60 GWh; it is located on the right river bank near the downstream dam toe [19].

The Uvac dam was built in 1979 on the Uvac River (at a distance of 73 km from the river mouth) with the idea of forming the Sjenica accumulation reservoir at a height of 989 m above sea level. The Uvac dam is the second largest embankment dam in Serbia; it is 307 m long and 110 m high, and has a shaft spillway reaching 30 m in diameter and a capacity of 1050 m^3/s. The catchment area of the Sjenica reservoir is 920 km^2; it has an average flow rate of 11.5 m^3/s and a total volume of 200×10^6 m^3. The derivation unit Uvac has one power generator with a capacity of 36 MW, an installed flow rate of 43 m^3/s, a calculated net flow of 97.5 m and an average annual production of 72 GWh [19]. It is located on the river bank at the end of the backwater of the Kokin Brod reservoir.

The Potpeć dam was built in 1967 on the Lim River (at a distance of 54 km from the river mouth), 8 km upstream from the town of Priboj. This is a gravity concrete dam, 218 m long and 46 m high, with a spillway capacity of 3000 m^3/s and a bottom outlet with a capacity of 240 m^3/s. The catchment area of the Potpeć reservoir is 3605 km^2; it has an average flow rate of 91.9 m^3/s, and a total volume of 27.5×10^6 m^3. The HPP Potpeć is a facility with an independent powerhouse with a capacity of 52 MW (3×17.5 MW), an installed flow rate of 165 m^3/s, a calculated net fall of 37.6 m and an average annual production of 200 GWh [19].

The Morava River Basin

The Morava River Basin includes the Velika Morava River and its tributaries: the Zapadna Morava and the Južna Morava. The Velika Morava River (182 km) is the most significant right-bank tributary of the Danube River. The river basin is placed almost entirely at the Serbian territory, and includes mid, southern and southeast parts of central Serbia. The total catchment area is 38,207 km^2, and 95.9% is located within Serbia. Zapadna Morava River is draining the southwestern part of Serbia with total catchment area of 15,696 km^2; 97.3% is located in Serbia while the rest is in Montenegro. Južna Morava River is draining the southeastern part of Serbia with total catchment area of 15,696 km^2; 92.7% is in Serbia and the rest is in Bulgaria [22]. Owing to favourable conditions for exploiting hydropower potential in the Morava River basin, several hydropower plants were built. The sub-basin of the Zapadna Morava includes facility Gazivode on the Ibar River. The HPP Pirot and the hydropower plants on the Vlasina River (Vrla I–IV) belong to the Južna Morava River basin.

The Gazivode dam was built in 1977 on the Ibar River, in the municipality of Zubin Potok in northern Kosovo and Metohija, near the village of Gazivode. It was put into operation in 1981. This is an earth-fill embankment dam with a clay core; it is 520 m long 108 m high. The normal backwater elevation of the Gazivode reservoir is 693 m above sea level; the average flow rate is 14.5 m^3/s, and the total volume is 370×10^6 m^3. This is a multi-purpose reservoir, primarily used for water supply to Kosovska Mitrovica, the Obilić thermal power plant and the irrigation of the Kosovo plain. The HPP Gazivode has a capacity of 35 MW, an installed flow rate of 42.6 m^3/s, a calculated net fall of 87 m and an average annual production of 70 GWh [8].

The dam on the Visočica River, which belongs to the basin of the Južna Morava River, was built in 1989 near the village of Zavoj, at the place where in 1963 a landslide created a barrier in the Visočica River, leading to the formation of Zavojsko Lake. After nearly two years, in December 1964, a tunnel was dug; its entrance was mined near the dam and the water from the lake flew into the Visočica River. The same year, plans were devised for the construction of an reservoir at the same place and a hydropower plant near Pirot. The plant was put into operation in 1990. According to data provided by the Đerdap Hydropower Plant company, this is an embankment earth-fill dam, 262 m long and 86 m high; it has three spillways with a capacity of 1820 m^3/s. The normal backwater elevation of the Zavoj reservoir is 616 m above sea level; it has an average flow rate of 3.08 m^3/s and a total volume of 170×10^6 m^3. The water from the reservoir is taken to a surge tank via a tunnel (9093 m long, with a diameter of 4.5 m), to be further supplied to a powerhouse through a two-tube steel pipeline. The HPP Pirot, with a power of 80 MW (2×40 MW), an installed flow rate of 45 m^3/s, and a calculated net fall of 206 m, belongs to peak-power plants because it operates 4–5 hours a day (about 1,400 hours per year), i.e., during the peak daily electricity consumption, and produces up to 75 GWh [17]. The HPP Pirot is used by the Electric Power Industry of Serbia, namely, the Đerdap Hydropower Plant Joint-Stock Company.

Another important system in the basin of the Južna Morava River is the Vlasina hydropower system, which includes Vlasina Lake, four hydropower plants on the Vrla River and the pumped storage unit Lisina. Vlasina Lake is the oldest reservoir in Serbia; it was built in 1949. The lake was created at a height of 1214 meters above sea level by the construction of a 34-metre high and 239-metre long earth-fill embankment dam on the Vlasina River. The volume of the lake is 176×10^6 m^3 and its water is used for hydropower production and water supply. This is also the highest lake in Serbia. At lower water levels, the water from Vlasina Lake is taken to an intake structure (which is under water at higher water levels) by an open channel; through a 2139-metre long tunnel and 1325-metre long pipeline, it is further supplied to four turbines located in the underground powerhouse of the HPP Vrla I. The first two power generators of the HPP Vrla I were put into operation in 1955, while another two were set to work in 1975. The HPP Vrla I has a capacity of 51 MW, an installed flow rate of 18.3 m^3/s, a calculated net fall of 338 m and an average annual production of 95 GWh. By processing one cubic metre of water from Vlasina Lake, the power generators of the Vrla I power plant produce 0.74 kWh; this is the greatest production per a cubic metre of water in the system. The processing through the remaining three steps yields an overall effect of 1.858 kWh per a cubic metre of reservoir water [17]. The HPP Vrla II is also located in the river bed of the Vrla River, at the mouth of the Bitvrđa River. The reservoir of the Vrla II power plant, located immediately below the output waters of the Vrla I power plant, has a volume of 0.1×10^6 m^3. It was formed by the construction of a dam in the river bed. The dam is 23 m high and 80.5 m long. The HPP Vrla II has a capacity of 24 MW, an installed flow rate of 18.5 m^3/s, a calculated net fall of 152 m and an energy–water ratio of 0.334 kWh/m^3, resulting in an average annual production of 51 GWh. Two power generation units were put into operation in 1954 and 1975. The HPP Vrla III is located in the village of Masurica, on the right bank of the eponymous river. The water from the reservoir (total volume: 0.05×10^6 m^3) is taken via a 7333-metre long tunnel into a surge tank and further, by means of a pipeline, to the powerhouse. Two power generators of the HPP Vrla III have a total capacity of 29 MW, an installed flow rate of 18.5 m^3/s, a calculated net fall of 185 m, an energy–water ratio of 0.44 kWh/m^3 and an average annual production of 73 GWh. The Vrla III power plant is the core of the Vlasina system where the operation of all four plants and the entire Vlasina system are monitored. The last, fourth step–the Vrla IV hydropower plant, is located by the Preševo–Niš–Belgrade highway near Vladičin Han. The HPP Vrla IV has a total capacity of 25 MW, an installed flow rate of 18.5 m^3/s, a calculated net fall of 164 m and an energy–water ratio of 0.344 kWh/m^3 and an annual production of 63 GWh [17].

3.2. Overview of Small Hydropower Plants in Serbia

In this section, we will present an overview of small hydropower plants smaller than 10 MW, drawing attention to their performance and other characteristics. The existing small hydropower plants are mostly located in western and southeastern part of Serbia (Figure 5).

The idea of building hydropower plants in the Ovčar-Kablar gorge dates back to the period after 1896, when a catastrophic flood of the Zapadna Morava at the entrance into the gorge caused landslides on the right bank which formed a dam, resulting in a considerable fall in a rather short river course. If a small dam had been built and a tunnel had been dug, the necessary conditions for a hydropower plant would have been provided. After the end of World War II, the concept of using the potential of the Zapadna Morava in the Ovčar-Kablar gorge was expanded to include the entire area. As a result, a system consisting of a derivation unit in Ovčar Banja and a downstream plant with an independent powerhouse in Međuvršje was constructed [20]. The Ovčar Banja dam was built in 1954 on the Zapadna Morava River, at the entrance into the Ovčar-Kablar gorge upstream from the village of Ovčar Banja. According to a technical report of the Drina-Lim Hydropower Plants company, this is a concrete gravity dam, 45 m long and 12 m high. The maximum backwater elevation of the Ovčar Banja reservoir is 295.2 m above sea level [23]; the average flow is 34.4 m^3/s, while the total volume is 3×10^6 m^3. From the reservoir, water is taken through a tunnel into the underground hydropower plant Ovčar Banja; it has a capacity of 6 MW and an installed flow of 40.0 m^3/s and a calculated net fall of 20 m.

The Međuvršje dam was built in 1957, on the Zapadna Morava (at a distance of 183 km from the river mouth), at the exit from the Ovčar-Kablar gorge near the village of Međuvršje. This is a concrete gravity dam, 189 m long and 32 m high. The normal backwater elevation of the Međuvršje reservoir is 274 m above sea level [23]. According to a technical report of the Drina-Lim Hydropower Plants company, the reservoir has an average flow rate of 34.4 m^3/s and a total volume of 18×10^6 m^3. The HPP Međuvršje is a facility with an independent powerhouse with a capacity of 7 MW, an installed flow rate of 40.0 m^3/s and a calculated net fall of 19 m. The total capacity of the Ovčar Banja and Međuvršje hydropower plants was 13 MW. Annually, they produced about 60 GWh [20]. The Ovčar Banja and Međuvršje hydropower plants are used by the Electric Power Industry of Serbia, namely the joint-stock company Drina-Lim Hydropower Plants Joint Stock Company.

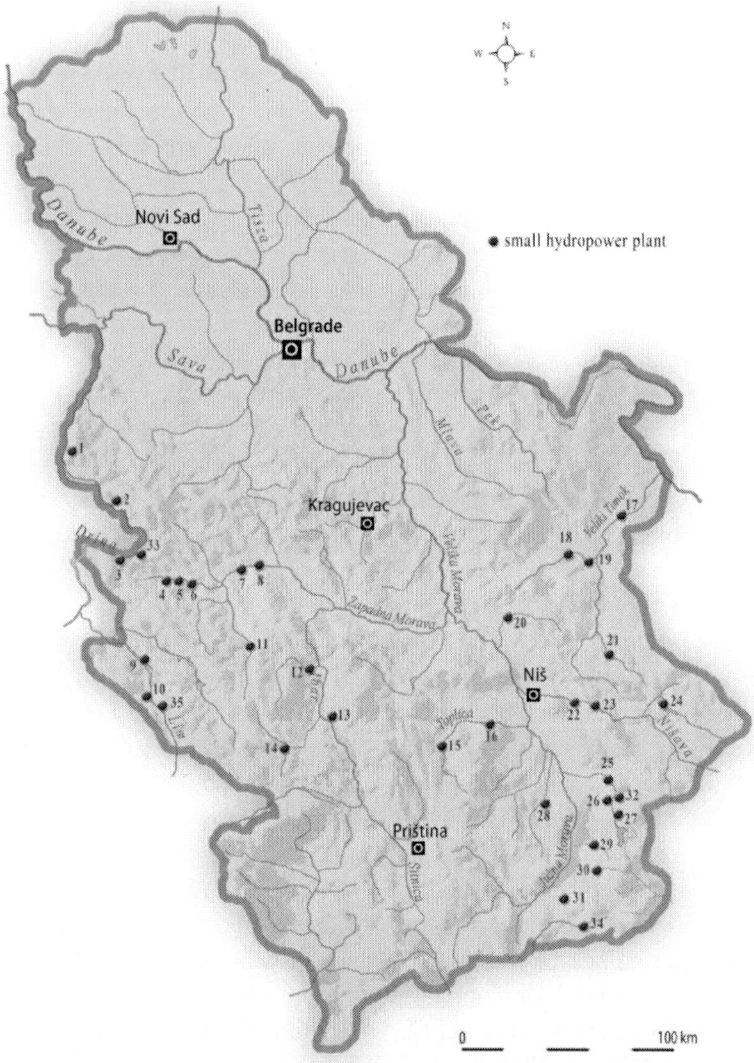

Figure 5. Small hydropower plants in Serbia (also see Table 3).

The SHP Prvonek is located at the foot of the Prvonek dam on Banjska River. The Prvonek reservoir, with a total capacity of 23×10^6 m^3, serves as a drinking water supply for four municipalities in south part of Serbia: Vranjska Banja, Vranje, Bujanovac and Preševo. This facility was constructed during the 2011 and was inaugurated in 2012 and it is first electric power facility built in compliance with the Decree on the Implementation Programme of the 2015

Energy Development Strategy, namely, with its provisions related to the construction of new electric power plants within the existing water management facilities. Also, it is the first electric power facility built by the Electric Power Industry of Serbia (EPS) in the past 20 years. The funds for the construction of the SHP were 1.6 million euro. The SHP has two hydro-aggregates: a small one, which employs reserved flow for electric power production, and the larger one, which uses available surplus water from the reservoir for energy production. Water is supplied to the small hydro-aggregate via a short steel pipeline with a diameter of 300 mm. The large hydro aggregate is supplied with water through a 145 m long steel pipeline with a diameter of 800 mm. The plant is connected to a 10-kV electricity distribution network through a 10/0.4 kV transformer station placed in the machinery hall. The operation control system of this small hydropower plant is integrated with the operation control system of the reservoir; accordingly, the owner (user) of the dam controls the water release regime, including the water release through the hydro aggregates within the SHP. Both aggregates are equipped with horizontal-axis Francis turbines. The large hydro-aggregate has the net head of 76 m and the installed discharge of 1.2 m^3/s; whereas the small hydro-aggregate has the net head of 50 m and the installed discharge of 0.25 m^3/s. Synchronous generator parameters are the same for both hydro-aggregates: the voltage of 0.4 kV and the frequency of 50 Hz. The total installed power is 0.91 MW, whereas the expected annual production is 3 GWh [24].

The SHP Vrelo is located in village of Perućac on the Vrelo River in western Serbia, near border with Bosnia and Herzegovina. Total length of the Vrelo River is 365m with mean annual discharge of 2.1 m^3/s. This SHP was built 1927 and it operated until 1959, when ti was closed due to frequent breakdowns and deterioration. In period 1985-1986, the equipment was fully repaired and upgraded. A new generating unit with a Francis turbine, manufactured by Ganc Danubius Budapest, and an asynchronous generator, manufactured by Sever Subotica, was installed. The total installed power was 90 kW. In 1986 this SHP was connected to distribution network. Today, due to deterioration of equipment SHP Vrelo works with installed power of 0.03 MW, average annual production is 0.25 GWh. For period 1986-2011 SHP Vrelo produced 4.8 GWh. A specific feature and an advantage of the SHP Vrelo is its vicinity to the HPP Bajina Bašta. Due to this, the same workers are engaged in its maintenance [22, 24].

4. Modernization and Upgrading of Existing Hydropower Plants

According to strategic document related to the future development of the energy sector in Serbia, one of the main goals is to increase the share of renewable energy sources in energy production. Having this in mind and taking into account the age and condition of the equipment at hydropower plants, the rehabilitation and upgrading of the existing large and small hydropower plants are also priority issues. As much as 63% of the installed capacity of hydropower plants in Serbia is older than forty years, while as many as six hydropower plants with a total capacity of 362 MW (13% of the total installed capacity) are older than fifty years. Having in mind that all hydropower plants larger then 10MW and a half of the total number of SHPs in Serbia are owned by the state, which manages them through the Public Company Electric Power Industry of Serbia, the overall objective of upgrading hydropower plants is to prolong their life of operation and ensure a reliable electric power supply under economically affordable and environmentally acceptable conditions in the future.

The specific objectives of upgrading and rehabilitation are to increase the availability and reliability of power generating units, increase the capacity and energy production, reduce losses, utilize overflow water and ensure a more economical and easier maintenance. It is estimated that the upgrading and rehabilitation of existing hydropower plants would increase their output to 128 MW (i.e., by 4.5%) and their annual production to 247.5 GWh (i.e., by 2.3%) [25]. In order to achieve the planned strategic objectives in the energy sector, in the previous period ending 2013, the rehabilitation of the HPP Ovčar Banja, HPP Međuvršje and HPP Bajina Bašta was completed, while the rehabilitation of the HPP Đerdap I was begun. The rehabilitation of the HPP Zvornik is about to begin. Furthermore, the rehabilitation of the Vlasina hydropower plants, as well as the Bistrica and Potpeć plants and the reversible HPP Bajina Bašta has been planned as part of the scheme aimed at increasing energy efficiency.

The rehabilitation of the hydropower plant system on the Zapadna Morava (the Ovčar Banja and Međuvršje power plants) was carried out between 2008 and 2010. According to data provided by the Drina-Lim Hydropower Plants Company, during the rehabilitation project, about 95% of the electromechanical equipment of turbines and power generation units were replaced, while other equipment was fully replaced. Turbines and generators were repaired by ATB Sever Subotica in collaboration with the France-based Andino Company. Before the rehabilitation, the HPP Ovčar Banja had a capacity of 6 MW; after the refurbishment, the turbine power has been 8.26 MW (3.2 MW and 5.06 MW). The run-of-river power plant Međuvršje had a capacity of 7 MW before the rehabilitation; after the refurbishment, the installed capacity of turbines has been 9 MW (3.4 MW and 5.6 MW). Due to the rehabilitation project, the total capacity of both hydropower plants has been increased by 25% (to 17 MW); accordingly, electric power production was also increased.

Late in 2013, the rehabilitation of the HPP Bajina Bašta was completed. During a four-year period, all of the four power generation units were refurbished. The replaced equipment amounted to 1200 t; four new turbines manufactured in Ravensburg, Germany, were installed, as well as new generators, which were manufactured in Weiz, Austria by Andritz hydro; all of the four transformers were replaced with new units and the complete high-voltage network was refurbished [26]. The local command system was replaced with modern interconnected master units. As a result, the capacity of the hydropower plant was increased from 364 MW (4×91 MW) to 420 MW (4×105 MW) [24]. Power generators were introduced into regular operation late in 2013, which increased the plant's capacity to 56 MW and led to an increase in electric power production by about 40 GWh per year. Due to the refurbishment, the power generator's life of operation has been prolonged by 35–40 years. At the HPP Bajina Bašta, the construction of the fifth generator unit has been planned.

The HPP Đerdap I has so far produced as much as 239,000 GWh of electric power, i.e., more than 5,500 GWh per year. In November 2013, an annex to the agreement between the Russian company Silovie Machini, the Public Company Electric Power Industry of Serbia and the Đerdap Hydropower Plant Company on the rehabilitation of the HPP Đerdap I was signed. Owing to this agreement, the life of operation of the plant should be prolonged by forty years, while the plant's capacity should increase by 66 MW.

According to the preliminary rehabilitation and upgrading project, the same rotating runner diameter of 9.5 m is to be maintained, the flow rate should be increased from 800 to 840 m^3/s, the capacity should rise from 194 MW to 205 MW, i.e., to 210 MW for a limited operating period of 100 hours per year, and the production of electric power in six generator units is expected to increases by about 80 GWh/year. The total value of the works to be implemented at the HPP Đerdap is expected to be 216.57 million dollars; out of this amount, 100.51 million dollars should be provided from the clearing debt of the Russian Federation, while the remaining 116.06 million dollars are to be allocated from the investment funds of the HPP Đerdap [27]. The rehabilitation of the HPP Đerdap began in 2009; the power generation units A6 and A4 have so far been upgraded. Since November 2013, the power generator A5 is under reconstruction and the work will continue throughout 2014. During the rehabilitation, the main equipment and a part of the auxiliary power generation facilities, such as the turbine runner, the blades and the upper variation ring, the generator stator, the rotor poles, the command system, the block transformer and multiple pieces of auxiliary equipment, were replaced with new devices [26]. Just like in other power generators, the installation of new turbine runner blades in the A5 unit will result in a greater efficiency and; accordingly, the power generation capacity is expected to increase by about 10% [17]. The full rehabilitation of the power plant is due to be completed in 2018.

The rehabilitation of the HPP Zvornik plant is expected to begin in 2015 and during a five-year period all of the four power generators should be replaced with new units. According to data provided by the HPP Zvornik, this would increase the capacity of the plant by about 30% (from the current 96 MW to 125.6 MW), while electric power production would rise by about 15% per year. According to the project, the entire turbine and turbine control equipment, cooling water and compressed air systems, power generators, block transformers and generator voltage equipment are to be replaced with new devices; hydromechanical equipment should be rehabilitated and the system for heating, cooling and ventilation of powerhouses should be reconstructed. The plan also foresees the installation of a modern control command system; furthermore, oils and fats should be eliminated from the runner boss and turbine bearings in order to make the system environmentally friendly; this would prevent possible pollution of the Drina River.

5. Future Hydropower Development Projects

In order to increase the production capacity in electric power generation from hydropower, some preparatory activities (projects, investment and technical documentation, environmental risk assessment studies, spatial plans, urban planning documentation) have been undertaken; they are aimed at building new production facilities in collaboration with potential partners. The strategies and projects of the Ministry of Energy, Development and

Environmental Protection deal primarily with the exploitation of those rivers that may be of interest for power production. Accordingly, the possibility of building new hydropower plants has been considered within the following projects: Project HPP Ibar; Project HPP Morava; Project HPP Srednja Drina; Project Reversible HPP Đerdap III, Project Reversible HPP Bistrica (Table 4 and Figure 2).

The Đerdap III Reversible Hydropower Plant Project is aimed at building a pumped-storage system that would operate as a pump station over night, while during daytime, i.e., during peak power consumption periods it would serve as a hydropower plant. The concept of the project implies the utilization of the existing reservoir of the HPP Đerdap I as the lower pool and the construction of an upper pool in the valleys of the Pesača (Phases I and II) and Brodica (Phases III and IV) rivers. The total capacity of the new facility would be about 1800 MW and it would produce about 460 GWh of electric power per season. The project is to be carried out between 2016 and 2020.

According to the project, the capacity of the Reversible HPP Bistrica will be 680 MW and it will be located near the existing HPP Bistrica. Regulated water from the already existing upstream reservoirs (Uvac and Kokin Brod) should be used to move the turbines in this new facility, as well as in the existing HPP Bistrica. A feasibility analysis of the project has been underway, whereas the construction is expected to be completed between 2015 and 2020.

The general design for the Velika Morava hydropower plant and previous studies, show that the most appropriate solution would be the construction of several cascading hydropower plants with independent powerhouses (HPP Varvarin, HPP Mijatovac, HPP Svilajnac, HPP Trnovče and HPP Ljubičevo), a total installed capacity of approximately 147.7 MW and an annual production of around 645.5 GWh. The construction should be carried out between 2016 and 2021.

A general design and a feasibility study for the HPP Ibar were drafted in 2010. The concept of the exploitation of the river's hydropower potential involves the construction of ten cascading hydropower plants with independent powerhouses (HPP Bojanići, HPP, Gokčanica, HPP Ušće, HPP Glavica, HPP Cerje, HPP Gradina, HPP Bela Glava, HPP Dobre Strane, HPP Maglič, HPP Lakat), a total installed power of about 103 MW and an annual production of 418.6 GWh. The construction should be carried out between 2014 and 2021.

Table 4. Priority projects aimed at the construction of new hydropower plants larger than 10 MW [28]

Project	River	Pi (MW)	W (GWh/year)	HPP type	Investment (million €)	Project duration
HPP Bistrica	Uvac/Lim	680	/	reversible	600	2015–2020
HPP Djerdap III*	Dunav	600	/	reversible	400	2016–2020
HPP Ibar	Ibar	103	419	run-of-river	300	2014–2021
HPP Srednja Drina	Drina	321	1197	run-of-river	819	2014–2023
HPP Velika Morava	V. Morava	148	645	run-of-river	250	2016–2021

*the first phase

The Srednja Drina Hydropower Plant Project (to be implemented between 2014 and 2023) has a strategic significance. The project should involve the construction of three hydropower plants with independent powerhouses (HPP Rogačica, HPP Tegare and HPP Dubravica), a total installed capacity of 321 MW and an annual production of 1197 GWh.

Projects aimed at the construction of small hydropower plants are constantly under consideration. It is believed that a significant number of approved projects will be completed by 2020. The facilities that are now being constructed are expected to have a total installed capacity of about 387 MW: 63 permits for the installed capacity of about 285 MW and 128 approvals (nominal capacity less than 1 MW) for the installed capacity of about 102 MW have been issued [25]. It is also planned to revise the *Survey of Small Hydropower Plants in the Republic of Serbia* (without the autonomous provinces) and the *Survey of Small Hydropower Plants in Vojvodina* but until that, these documents remain the key documents concerning the future development and construction of SHPs in Serbia. According to these documents the greatest hydro potential usable for SHPs is located in the west (in the Kraljevo and Užice regions) and south (the Niš region) parts of Serbia, which are mountainous areas. The greatest number of such facilities could be installed in the region where the first small hydropower plant in Serbia was built–in the Užice region. In the northern, flatland area of Serbia (Podunavlje, Vojvodina, the Belgrade area), hydropower potential is somewhat lower but, having in mind that these regions are the most densely populated and economically most developed parts of Serbia, it should also be taken into account [25].

The utilization of the hydropower potential of municipal water supply dams also has an important role. In that respect, the Spatial Plan of the Republic of Serbia (2010-2020) foresees the drafting of the investment-related and technical documentation and the implementation of projects aimed at installing SHPs in the existing accumulation dams and hydropower production facilities (SHPs Jezero, Mala Vrla, Zavoj, Pirot, Nikola Tesla, Banjica, etc.), as well as in the existing multi-purpose water management accumulation reservoirs (Ćelije, Bovan, Barje, Parmenac, Grlište, Vrutci, etc.). The construction of new accumulation reservoirs for SHPs (on the rivers of Veliki Rzav, Studenica, Jerma, etc.) has also been planned. According to the plan, the capacity of newly constructed SHPs should not be smaller than 45 MW. Along with the existing dams within the Danube–Tisza–Danube hydro-system, the construction of ten SHPs (with a power between 0.1 and 10 MW) has also been foreseen. The construction of such facilities would significantly increase the flow capacity of the canal network, which is expected to have an environmentally beneficial effect.

Furthermore, a possible line of development would be the reconstruction and adaptation of the sites where water-mills were constructed in the past (according to estimations, there are about 5000 such locations in Serbia). It is estimated that this would enable to generate 10 MW of the installed power, i.e., about 45 GWh of electric power per year. However, technical documentation for these facilities is lacking and it would be very difficult to assess investment possibilities [25].

6. Environmental Issues

Hydropower plants rest on a form of energy generation where there are hardly any emissions of pollutants – neither in the atmosphere nor into the water bodies. In the Danube Declaration (2010), Danube countries highlighted that "*hydropower plants offer an additional reduction potential for greenhouse gases but recognizing as well their negative impacts on the riverine ecology*" [29, p.58]. Apart from the numerous positive aspects of hydropower, it can also create serious social and environmental problems (Table 5). The table shows the advantages and disadvantages of hydropower projects defined by the International Hydropower Association (2003).

The issues of the modes and nature of the impact (hydrological, geomorphological, biological, etc.), as well as the intensity of the impact of damming rivers and the construction of hydropower facilities (dams and reservoirs) on the environment are the subject of numerous scholarly papers. According to McCartney [31, p.122] "*in the last three decades, an increase in environmental awareness has led to the recognition that the management of water resources includes a responsibility to protect the users of water (and the natural resources that depend on water) from over utilisation or impacts that cause degradation*". As shown in Table 5, hydropower facilities (their construction and operation) cause a number of environmental impacts. They depend on the type of hydropower plants (run-of-river hydropower scheme; storage hydropower schemes and pumped storage hydropower schemes).

According to Richter and Thomas [32] run-of river facilities result in only minor alterations to flow regimes. Kumar and Katoch [33] have analyzed studies by various authors who highlight the environmental advantages of run-of-river facilities (RoR) compared to "storage-based projects"; the most typical attitudes include: river is not transformed into a lake; flow pattern of the river remains essentially unchanged which reduces downstream geomorphological and biological impacts of the project (e.g., fish migration is less affected); almost always less costly as compared to storage plants. However, the fact that RoR hydropower projects have some environmental limitations should not be neglected. As far as the ecological consequences of hydropower dams with storage capabilities are concerned, they are related to the elimination of small floods, introduction of frequent, artificial high-flow pulses, and the lowering of river levels below natural low-flow levels Furthermore, the construction of hydropower facilities required high capital cost, rehabilitation, resettlement of people, and geographical disturbance. Pumped-storage systems also have less severe environmental impacts compared to storage-based projects, because in a pumped-storage system less water needs to be stored in reservoirs and much of the natural inflows into the hydropower reservoir can be passed downstream, thereby moderating impacts on natural flow patterns [32]. Comparing large and small hydropower plants, Lenzen [34] see small hydropower plants as environmentally benign and able to alleviate poverty and foster sustainable development, especially in developing countries. While large hydropower plants built over the past decades account for more than 90% of the world's current capacity, supplying about 17% of the world's electricity, they are also associated with major environmental and social impacts.

The construction of hydropower dams (damming rivers) and impoundment of water (forming reservoirs) are considered to be a form of *human alteration of natural flow regimes*, i.e., the spatial and temporal variations of their component parts, such as: the magnitude, frequency, duration, timing, and the rate of change [35]. The flow modification caused by damming a river is one of the five threats to global freshwater biodiversity, along with overexploitation; water pollution; destruction or degradation of habitat and invasion by exotic species [36]. By forming a reservoir, natural water regimes are transformed into managed water regimes [37]. Also, according to the Regulations on the identification of water bodies and groundwater (Official Gazette of RS, no. 96/10), water bodies in the rivers in Serbia are classified into rivers (natural water bodies), significantly modified water bodies and artificial water bodies. The changes that occur in those sections of watercourses where HPPs (dams and reservoirs) are built and those that occur in rivers (surface water) upstream from the reservoir backwater or from the HPP, i.e., the sector of the watercourse downstream from the dam where there are changes in the hydrological and sediment regimes, are the best

Table 5. Advantages and disadvantages of HPP [30]

Advantages	Disadvantages
Economic aspects	
Provides low operating and maintenance costs Provides long life span (50 to 100 years and more) Provides reliable service Includes proven technology Instigates and fosters regional development Provides highest energy efficiency rate Creates employment opportunities and saves fuel	High upfront investment Precipitation Requires long-term planning Requires long-term agreements Requires multidisciplinary involvement Often requires foreign contractors and funding
Social aspects	
Leaves water available for other uses Often provides flood protection May enhance navigation conditions Often enhances recreation Enhances accessibility of the territory and its resources Improves living conditions Sustains livelihoods (fresh water, food supply)	May involve resettlement May restrict navigation Local land use patterns will be modified Waterborne disease vectors may need to be checked Requires management of competing water uses
Environmental aspects	
Produces no pollutants but only very few GHG emissions Enhances air quality Produces no waste Avoids depleting non-renewable fuel resources	Inundation of terrestrial habitat Modification of hydrological regimes Modification of aquatic habitats Water quality needs to be managed
Often creates new freshwater ecosystems with increased productivity Enhances knowledge and improves management of valued species due to study results Helps to slow down climate change Neither consumes nor pollutes the water it uses for electricity generation purposes	Temporary introduction of methylmercury into the food chain needs to be monitored/managed Species activities and populations need to be monitored Barriers for fish migration, fish entrainment Sediment composition and transport may need to be monitored/managed

indicators of the "significantly modified water body" category. According to the Law on Waters (Official Gazette of RS, no. 30/10), this body is defined as a body of surface water whose characteristics are considerably changed due to physical changes caused by human activity. In compliance with the Water Framework Directive (EC) the Law on Waters (Official Gazette of RS, no. 30/10) requires that these water bodies reach good ecological potential instead of a good status, which is impossible to achieve given the hydro-morphological changes in the water body (Regulations on the parameters of the ecological and chemical status of surface waters and the parameters of the chemical and quantitative status of groundwater, Official Gazette of RS, no. 74/2011).

6.1. Diverse Impacts of Hydropower Plants on River Ecosystem

There are many studies dealing with the environmental/ecological impacts of hydropower dams or HPPs [38, 29, 39, 40, 31, 41, 42, 43]. According to WCD [43, p.74], ecosystem impacts are classified as: "*the first-order impacts* (physical, chemical, and geomorphological consequences of blocking a river and altering the natural distribution and timing of the streamflow); *the second-order impacts* (changes in the primary biological productivity of ecosystems, including effects on riverine and riparian plant-life and on downstream habitats, such as wetlands); and the *third-order impacts* (alterations to fauna,such as fish, caused by a first-order effect, such as blocking migration, or a second-order effect, such as a decrease in the availability of plankton)". According to McCartney [31], the impact of dams can be divided into the *impact on abiotic variables* (river flow, thermal regime, water chemistry/quality, sedimentation) and *the impact on biota* (primary productivity, molluscs, fish, birds and mammlas), while Mc Cartney, Sullivan and Acreman [40] divide the ecosystem impacts of large dams into *upstream impacts* and *downstream impacts*.

The environmental impacts of dams and reservoirs are numerous; the most significant ones include:

- *impacts on the hydrological regime*–changes in the distribution and magnitude of flows, i.e., changes in the timing, magnitude, and frequency of low and high flows [44, 45, 46, 47]; hydropeaking–unnatural flow variation [48]; hydropeaking in particular can affect fish through physical displacement, behavioral changes, and loss of habitat; effects on the regime of ground water [49].

- *sediment-related impacts* - changes of sediment yield, transport and load and impacts on the composition of sediments [50, 51, 52, 53, 54, 55, 56]; siltation reservoirs [57, 58, 59, 60]. The problem of sediment in dammed rivers is closely related to the overall situation in the basin, i.e., with numerous interacting variables [61].

Babić Mladenović, Obušković, & Knežević [57, pp.389-391] have presented an overview of reservoirs sedimentation in Serbia:

- The reservoir of the HPP Bajina Bašta - in the first analyses, the sediment input was estimated at 210 m^3/km^2/year (3.2×10^6 m^3/year). The construction of upstream reservoirs on the Drina River (Kokin Brod, Potpeć, Piva, Uvac and Višegrad) has led to changes in the water regime and has also had a certain impact on the retention of sediment in the reservoir of the HPP Bajina Bašta. During the first ten years of operation (1966–1976) the precipitated sediment amounted to 25×10^6 m^3. The second survey of the storage (1989) showed that the precipitation in the reservoir slowed down. Based on the extrapolation of available data it was possible to estimate that by 2002, the total volume of the reservoir was reduced from the initial 341.8×10^6 m^3 to 283×10^6 m^3. The volume of dead space was reduced by 28.9%, while the storage capacity was reduced from 205.4 to 186 $\times10^6$ m^3. Based on these data, a preliminarily conclusion was that the process of siltation in the reservoir was not intensive; accordingly no erosion protection measures were undertaken
- The reservoir of the HPP Potpeć–after the dam had been constructed (1967), a "fairly" rapid siltation of the reservoir was observed. Until 1975, there were intense water level fluctuations in the reservoir, whereas in later periods the operation mode has involved the maintaining of a high and stable upper-dam level during periods of high water levels. The reservoir was emptied on several occasions (in 1979 and in 1980, this caused the withdrawal of large quantities of sludge, severely affecting the fish stock and causing a fish kill in the downstream section of the river; the reservoir was also emptied in 1988 and 1999). It is estimated that 350 m^3/km^2/year of sediment reach the reservoir from the Lim River basin. Erosion protection works aimed at reducing the quantity of sediment by up to 80% have

not been implemented. Based on the extrapolation of available data, it has been calculated that during a 35-year exploitation period, the total volume of the reservoir decreased from 42.9 to 24.4×10^6 m^3; the sediment was mostly deposited in the dead area (reduced by 15.2×10^6 m^3 or 53.5%). The siltation of the useful area (reduced by 3.3×10^6 m^3) was considerably less pronounced (22.7%). A trend of reduced siltation of the reservoir has also been observed: from 250 m^3/km^2/year in the first period of operation, to 100 m^3/km^2/year in the recent period. It is intended to maintain a total volume of 20×10^6 m^3 and a useful volume of 10×10^6 m^3, which ensures flexible daily and even long-term water-level adjustments and mitigation of flood waves.

- The reservoir of the HPPs Kokin Brod and Uvac–at the time when these reservoirs were designed and after their construction (Kokin Brod, 1962; Uvac, 1979), little attention was paid to siltation. The mode of operation of these reservoirs is such that the entire amount of deposit remains in the reservoir (negligible amounts pass through turbines). The input of sediment is small, compared to the total (and "dead") volume. It has been estimated that only about 3% of the usable capacity have so far been lost. Based on available data, it has been concluded that the reservoirs of Kokin Brod and Uvac do not have and will not have a problem of siltation any time soon;
- The reservoir of the HPPs Ovčar Banja and Međuvršje–due to the rather small useful volume (shallow reservoirs), daily or weekly water-level adjustments were from the outset foreseen by the project. Rapid siltation Ovčar Banja reservoir began soon after the construction of the dams had been completed (1954), and Ovčar Banja reservoir lost as much as 50% of its capacity. Sediment flushing and sediment discharge caused by lowering water levels at the time of high water flows were performed in 1955/56. The entire amount of sediment was transferred into the Međuvršje reservoir, where it has remained. It is estimated that the process of siltation in the Ovčar Banja reservoir is completed (the total volume decreased by 86%, the "dead area" by 89%, while the volume useful volume decreased by 79.1%). The siltation process in the Međuvršje reservoir is almost complete (the total volume decreased by 82%, while the useful area decreased by 62%). The movement of sediment from the upper to the lower section has also been observed. It is estimated that the available useful volume of the Međuvršje, amounting to 1.0×10^6 m^3, will be

maintained in the long-term while in the Ovčar Banja accumulation reservoir, the useful volume is limited to 0.25×10^6 m^3

- The Đerdap Hydropower Plant - between 1972 and 2009 the input of suspended load into the Iron Gate 1 (IG1) reservoir by the Danube and its major tributaries was 619.3 million tonnes (or approximately 16.3 million tonnes per year). About 103.5 million tonnes of sediment (or 2.7 million tonnes per year) have passed through the turbines and over the spillways of the IG1 and Iron Gate 2 (IG2) dams. The IG1 reservoir retained approximately 452.2 million tonnes of sediment (or 12.6 million tonnes per year on average). The greatest amounts of sediment were deposited within the reservoir during the first three years. Approximately 77% of the incoming sediment was deposited within the IG1 reservoir during the entire period of operation. Along the downstream 160 km-long stretch of the IG1 reservoir, some 434.9 million tonnes have been deposited. In the "immediate IG1 reservoir" (between the IG1 dam and km 1075 of the Danube), the Danube's natural regime was completely altered as soon as the reservoir was filled and an sediment deposition started. Riverbed surveys indicated that in the period between 1976 and 1997, the highest sediment deposits were formed between km 970 and km 1003 of the Danube River (the Donji Milanovac zone). Surveys of the riverbed conducted after the 2006 flood and in 2010 indicate that the sedimentation zone is now shifting further downstream (between the dam and km 970). The area between km 970 and km 1003 cannot retain sediment anymore because it has reached a new equilibrium. The IG2 reservoir (considerably smaller than the upstream one) is almost without impounding effects [62], for greater detail cf. [51].

The examples of the reservoirs of the HPPs Međuvršje and Ovčar Banja on the Zapadna Morava and HPP Zvornik on the Drina, which lost much of their capacity due to siltation, are representative. The siltation of the reservoir of the HPP Đerdap I is specific due to large amounts of sediment contributed by the Danube River. Babić Mladenović, Obušković and Knežević [57] point out that the siltation regimes and dynamics in reservoirs vary, just like the situation in reservoirs. The general conclusion of the authors is that the problem of siltation in reservoirs has not been paid sufficient attention.

- *impacts on river geomorphology*–changes in the hydrological regime and sediment transport a watercourse due to damming (impoundment dams) directly affect river geomorphology, both upstream and downstream [63, 64, 65]; the reduction of sediment in rivers downstream from the dam results in an increased erosion of riverbanks and river beds, the loss of floodplains (through erosion and a decreased over-bank accretion) and degradation of coastal deltas [31]. Morphological alteration Danubes channel in Serbia are not only result of natural river processess, thy are also reflect impacts numerous factors (regulation works, excavation of material from riverbed and riverbanks, etc.) and, in particular, the boulding the Iron Gate Hydropower. The result of the channel survey are showing that due influence of Iron Gate Hydropower backwater, channel patterns of downstream stretches are severely modifide. Short stretch downstream of the Iron Gate II is considered moderately modified. Along the Iron Gate reservoir, former floodplains are now permanently below the water surface [66].
- *impacts on water quality*–the construction of dams causes a transformation from "rivers" to "lakes"; this may lead to changes in the physical, chemical and biological characteristics [67], degradation of water quality in a reservoir [68, 69, 70] and changes in the thermal regime[71]; the process of thermopeaking (short-term temperature fluctuations in the river reach below dams) is causally linked to hydropeaking [72]; ice jam formation on rivers [73]. As Babić Mladenović et al. [62] point out forest of the deterioritation of water quality of Iron Gate I (IG1) reservoir due to high water levels and low velocites reducing the streams selfpurification potential resulting in increased eutrofication effects as well as acumulation of hazardous and harmful substances in sediment [69], has not been fulfilled. According to all physical and chemical parameters, the water quality of IG I reservoir corresponds to the specified class II according to Serbian classification. Water quality continuously improves along Serbian sector of the Danube river [74]. The IG1 reservoir has an important role in this process, functioning as a *large physical and biochemical reactor where intense decomposition of organic matter and removal of organic matter and nutrients from the system takes place*. According to the biological parameters (phytoplankton and zooplankton), the water quality corresponds to class II, while macroinvertebrate species indicate class III water quality. There is no

temperature stratification and stable oxygen stratification, and there is a lack of an intensive eutrophication potential. The water quality is still satisfactory, due to the high capability of self-purification [62]. The results of the chemical analyses showed that the surface water in other Serbian natural lakes and artificial reservoirs is water quality of class II. However, the sample from reservoir Međuvršje, where the NH_4^+ concentration was 0.28 mg/l, and the sample from Ovčar Banja, where the total phosphorus (TP) content was 0.15 mg/l with a high total nitrogen (TN) content of 1.21 mg/l, which may indicate that the water contains harmful bacteria or other pollutants [68].

- *biological implications by river fragmentation/habitat alterations*–the modification of hydrological and geomorphological processes of stream and river systems induces ecological responses [75] and a disruption of connectivity as an important component of nearly all aspects of a functioning riverine system; various *impacts* are exerted on riverine ecosystems [76], biodiversity [77], riparian ecosystems [78, 79], stream fishes [80, 81], macroinvertebrate communities [82, 83], the zoobenthic community [84], etc.

The analysis of the above-mentioned studies leads to the conclusion that the relations in a river basin and a watercourse which arise due to a HPP (a dam and a reservoir) are complex, multiple and diverse and need to be tackled in an integrative manner [85, 86].

6.2. Measures to Mitigate Environmental Impacts Resulting from the Constructed Hydropower Plants and Environmental Protection Measures in the Design and Construction of New Hydropower Plants

Measures to mitigate environmental impacts resulting from the construction of hydropower plants are diverse [31, 87], the most important ones include [85]: enabling fish migration; ensuring ecological flow; ensuring sediment transport; mitigating effects of artificial flow/water level fluctuations (hydropeaking) and compensatory measures. Ensuring fish migration and ecological flow were identified as priority measures at European level as well as in the Danube basin for the improvement and maintenance of ecological status.

1. Enabling fish migration

As it has already been mentioned, the utilization of hydropower for electric power production has the most severe negative impact on the flora and fauna of aquatic ecosystems. Of particular importance is the interruption of natural migratory paths of the fish fauna, as well as frequent major water level fluctuations in a reservoir (hydropeaking), which have additional effects on the ecosystem and a very negative effect on the reproduction of fish, particularly of those species that lay eggs on the coast. Due to sudden drops in water levels, spawn and fry remain dry and die. The well-known migratory sturgeon species (Danube sturgeon, Atlantic sturgeon, bastard sturgeon, starry sturgeon and beluga), which used to come upstream along the Danube to Hungary and Slovakia prior to the formation of the Đerdap reservoir, can no longer be found in the Đerdap reservoir because the HPP Đerdap I interrupts their migratory path [88].

In order to achieve a sustainable hydropower utilization, it is necessary to ensure conditions for unhindered migration and spawning of fish all along the watercourse in the form of "fish passage" or other technical solutions (e.g., by-pass streams, fish lifts etc.), [89, 85]. In modern practice, solutions to this problem have to be offered already when designing new facilities. The HPP Zvornik is the only hydropower plant in Serbia which has a fish path, which has been opening regularly since 2003. It enables to re-establish the communication between the reservoir and the downstream section of the Drina River; the construction of the dam interrupted the migratory path used by of certain types of fish when spawning. The path is about 160 m long, cascading, with four rest areas, i.e., deeper pools in which fish remain longer, while in the barriers between cascades there are openings through which fish can pass. Every year, in April, the fish path opens to enable fish from the lower course of the Drina River to move to Zvornik Lake and spawn.

2. Ensuring ecological flow or guaranteed ecological flow

Two key indicators of ecological stability in projects aimed at the construction of hydropower plants with reservoirs of various degrees of regulation are: (a) guaranteed ecological flow downstream from dams [37, 90], (b) the analysis of the dynamics of abiotic and biotic processes in a reservoir with the aim of understanding the environmental situation in the lake after a certain period of time [91]. The guaranteed ecological flow is a water discharge that must always ensure normal survival and development of the biocoenosis in a river biotope [37]. Therefore, this type of discharge is an ecological category [92]. The guaranteed ecological flow method has been devised for watercourses in this part of the Balkan Peninsula; accordingly, it is adjusted to the water regimes of rives in this region (for greater detail cf. [37, 90])

3. Control the siltation

Measures to control the siltation of reservoirs can be classified into three main categories [57]: reducing inputs of sediment into reservoirs, preventing deposition of sediment in reservoirs and removing deposited sediment. Given that the reservoir sedimentation is a serious off-site consequence of soil erosion [93], the first category must imply the reduction of the production of sediment in upstream parts of the catchment area, soil conservation, the control of erosion and torrent control [94, 95]. The second category of measures is based on the concept that it is easier to maintain sediment in suspension during its passage through the reservoir, than to move deposited sediment (which is especially difficult if sediment is consolidated). The third category includes the removal of sediments deposited at the bottom of a reservoir, creating hydraulic conditions suitable for the start-up and transport of sediment and excavated material by mechanical or hydraulic means [57].

4. Mitigating effects of artificial flow/water level fluctuations (hydropeaking)

Hydropeaking alters the natural flow regime with unsteady water release to rivers. Mitigating the adverse impacts of hydropeaking on aquatic ecosystems has become a crucial step in recent water policies. Mitigation measures for hydropeaking are subdivided into two concepts, namely operational (non-structural) measures and structural measures [96].

5. Monitoring

There is a general agreement concerning the importance of monitoring programmes for the study of the environmental impact of hydropower facilities (dams and reservoirs). Such programmes imply timely instrumentation data collecting and evaluation. The basis of any effective monitoring programme is a study design that allows for the detection of changes resulting from the operation of a hydropower facility. These principles underlie the multidisciplinary and complex programme for monitoring the environmental impacts and effects of the protection measures in the Đerdap Hydropower and Navigation System on the Danube River (HPNS), conducted by Jaroslav Černi Institute. The programme for the monitoring, measuring and analysis of the impact of the HPP Đerdap I (established in 1978) consists of nine specific sub-programmes for monitoring [62]: (I) the river flow and backwater levels; (II) the groundwater levels and the drainage system operation; (III) the sediment regime and deposition; (IV) the ice regime; (V) agricultural land preservation measures; (VI) forests and wetlands; (VII) flood control structures; (VIII) the quality of water and sediment; and (IX) the riverbank and landslide stability. In 1985, a very similar monitoring programme was set up for the Iron Gate 2 reservoir.

6. Strategic Impact Assessment / Environmental Impact Assessment

One of the most important preventive environment protection measures is to prepare a strategic impact assessment of environment-related policies, plans and programmes, as well as an impact assessment study of environment-related projects, including the construction of hydropower plants; such studies have become part of normal procedures both in worldwide and Serbia [97, 98, 99]. In Serbia, this measure is regulated by the Law on Environmental Impact Assessment (Official Gazette of RS, No.135/04, 36/09) and the Law on Strategic Impact Assessment (Official Gazette of RS, No.135/04 and 88/10). Assessing the environmental impact of a project is an integral part of the technical documentation; according to the procedure defined by a special law, the implementation of a project cannot be undertaken without an environmental impact study.

7. The application of mathematical models

As far as the planning and exploitation of hydropower systems are concerned, it is very important to consider and analyze the dynamic processes that will take place in reservoirs in the form of changes of abiotic and biotic conditions already during the designing phase. Therefore, in worldwide and Serbia, various mathematical models are being developed [91, 100, 101]. In order to enable an integrated management of the hydropower potential of Serbia, the Jaroslav Černi Institute for the Development of Water Resources has developed the following hydro-information systems: the Đerdap Hydro-Information System; the Drina Hydro-Information System and the Vlasina Hydro-Information System. The most important aims of such systems are to integrate all relevant hydrological, meteorological, hydrogeological, hydropower-related and other data, as well as to make them available to a wide circle of users; to make the best possible decisions related to the operational management of hydropower plants in various situations, as well as to make the best possible strategic decisions in choosing solutions for an integrated basin management [18]. These systems enable an optimum real-time management and the harmonization of various stakeholders.

CONCLUSION

Although the idea of installing hydropower plants in Serbia emerged already in the early 20th century, which was followed by the construction of several dozen such facilities, more recent political and economic changes caused a significant delay in their development, compared to other countries. Today, in accordance with a number of strategic and planning documents on the development of the energy sector in the Republic of Serbia, the aim is to increase the share of hydropower in total electric power production, using the latest technology in the use of renewable energy, increasing efficiency and economically justified investment.

One segment will be focused on the construction of large hydropower plants, while the other will deal with the construction of small hydropower plants. As the last major hydropower plant, HPP Pirot, was built in 1990, the construction of up 20 hydropower plants with a total installed power of 1852 MW in the next ten-year period is considered a priority issue. These power plants will be built in the river basins that have the greatest hydropower potential: on the Danube, Velika Morava, Drina, Ibar and Uvac.

As far as the further development of small hydro power plants is concerned, the crucial tasks is to revise the present *Survey of Small Hydropower Plants in the Republic of Serbia*, as well as to modify and adjust them to actual conditions and possibilities. According to the development strategy, until 2020, up to 63 SHPs (nominal capacity up to 10MW) with an installed capacity of approximately 285 MW and 128 SHPs (nominal capacity less than 1 MW) with an installed capacity of about 102 MW should be constructed. In addition to the planned new investments, the upgrading of the existing large and small hydropower plants should be undertaken in order to extend their life of operation and ensure the reliability of electric power supply.

In addition to the already planned investment projects, Serbia is committed to utilizing the hydropower potential of other watercourses where the construction of both large and small hydropower plants proves to be feasible. While determining the possibilities for the utilization of the remaining technical hydropower potential, non-energy criteria related to the multipurpose use of water and the political agreements on sharing hydropower potential with neighbouring countries should be taken into account.

On the other hand, the environmental aspect of building hydropower plants in Serbia should not be ignored. Construction of hydropower dams and reservoirs leads to significant hydro morphological changes in the water body, which as a result have an adverse impact on the ecosystems and habitats, such as breaking the continuity of the water, sediment and fish movements, changes in the morphology of the watercourse, the composition of the river bottom, changing the characteristics of the coastal area. Lessons learned about the effects of hydroelectric power plants on the environment provides a good basis for future hydropower projects in Serbia. Special attention must be paid to the problems of ensuring ecological flow, enabling fish migration and reservoir sedimentation. The upgrading of old plants implies the combination of technical upgrading with ecological restoration of existing hydropower installations [102].

REFERENCES

[1] Espinoza, L. J. & Barragán, A. E. (2013). Reneawable Energy Policy and Legitimacy: a Developing Country Case. *Renewable Energy and Power Quality Journal (RE&PQJ)*. International Conference on Renewable Energies and Power Quality (ICREPQ 13).

[2] Zimny, J., Michalak, P., Bielik S. & Szczotka, K. (2013). Directions in development of hydropower in the world, in Europe and Poland in the period 1995–2011. *Renewable and Sustainable Energy Reviews*, *21*, 117-130.

[3] Alonso-Tristán, C., González-Pena, D., Díez-Mediavilla, M., Rodríguez-Amigo, M. & García-Calderón T. (2011). Small hydropower plants in Spain: A case study. *Renewable and Sustainable Energy Reviews*, *15*(6), 2729-35.

[4] Kucukali, S. & Baris, K. (2009). Assessment of small hydropower (SHP) development in Turkey: Laws, regulations and EU policy perspective. *Energy Policy*, *37*(10), 3872-3879.

[5] Blyashko, Ya. I. (2010). Modern trends in the development of small hydropower around the world and in Russia. *Thermal Engineering*, *57*(11), 953-960.

[6] Ministry of Mining and Energy. (2009). *Energy Balance of Republic of Serbia for, 2008*, Belgrade.

[7] Tešić, M., Kiss, F. & Zavargo, Z. (2011). Renewable energy policy in the Republic of Serbia. *Renewable Energy*, *15*, 752-758.

[8] The Jaroslav Černi Institute, (2001). *Water Master Plan of Serbia.*

[9] Ministry of Mining and Energy of Republic of Serbia. *Decree on Incentive Measures for Privileged Electric Power Producers*. Official Gazette RS 8/2013.

[10] Jovanović, B. & Parović, M. (2009). *Stanje i razvoj malih hidroelektrana u Srbiji* (State and Developement of Small Hydropower Plants in Serbia; in Serbian*)*. Belgrade: Jefferson Institute.

[11] Golušin, M., Tešić, Z. & Ostojić, A. (2010). The analysis of the reneawble energy production sector in Serbia. *Renewable and Sustainable Energy Reviews*, *14*, 1477-1483.

[12] The Energy Law of the Republic of Serbia (Official Gazette of the Republic of Serbia 84/04, 57/2011).

[13] Statistical Office of the Republic of Serbia (2012). *Statistical Yearbook.* Serbia.

[14] Panić, M., Urošev, M., Milanović Pešić, A., Brankov, J. & Bjeljac Ž. (2013). Small hydropower plants in Serbia: Hydropower potential, Current State and Perspectives. *Renewable and Sustainable Energy Review*, *23*(6), 341-349.

[15] Working Group (2006). *Hidropotencijal opštine Pirot.* (Consideration of hydropower potential of Pirot municipality. Hydropower potential–small hydropower plants (SHPs) to 10 MW, In Serbian). Available from: http://www.pirot.rs/downloads/strategije/mhe_pirot.pdf. (Accessed June, 2012)

[16] Babić Mladenović, M., Divac Bartoš V. & Kolarov, V. (2010). *Natural characteristics of the Danube river in Serbia.* In The Danube in Serbia - The Results of National Program of the Second Joint Danube Survey 2007. Editors Simonović, P., Simić V., Simić S., Paunović M. Belgrade, 59-78.

[17] Privredno društvo 'Hidroelektrane Đerdap' d.o.o Kladovo, n.d. www.djerdap.rs (accessed June, 18 2014)

[18] Divac, D., Grujović, N., Milivojević, N., Stojanović, Z. & Simić, Z. (2009). Hydro-Information System and Management of Hydropower Resources in Serbia. *Journal of the Serbian Society for Computational Mechanics*, *3*(1), 1-37.

[19] Vučković, D., Melentijević, M., Milovanović, M. (2004). Postojeće stanje izgrađenosti na slivu Drine (The current state of development of the Drina River; In Serbian), *Vodoprivreda*, 207-208, 39-49.

[20] Drinsko Limske hidroelektrane Bajina Bašta.http://www.dlhe.rs (accessed June,18 2014).

[21] Milanović, A. & Kovačević, J. (2002). *Hydropotential of Drina river basin and its exploitation.* Proceedings from: International Scientific Conference in Memory of Prof. Dimitar Jaranov, *Development and State of Environment*, Varna, Bulgaria.

[22] Marković, D., Čeperković, B., Vlajčić, A. & Resl, S. (2011). *Bela knjiga elektroprivrede Srbije* (White book of EPS; In Serbian).

[23] Milanović A. (2000). *Vodne akumulacije za dobijanje električne energije u Srbiji* (Water reservoirs for electricity generation in Serbia; in Serbian), Graduate thesis, University of Belgrade, Faculty of Geography.

[24] The Jaroslav Černi Institute for the Development of Water Resources (2010/2011). *Aktivnosti i rezultati* (Activities and Results; In Serbian). http://www.jcerni.org (accessed June 2014).

[25] Ministry of Mining and Energy of Republic of Serbia (2007-2012). *Decree on Amendments and Supplement on the Program for the Realization of the Energy Development Strategy of the Republic of Serbia by 2015.* Official Gazette RS 72/09.

[26] The Electric Power Industry of Serbia (2013). Godišnji izveštaj (Annual Report, In Serbian).

[27] Ministry of Energy, Development and Environmental Protection of Republic of Serbia, n.d. http://www.merz.gov.rs/en (accessed June 2014)

[28] Ministry of Energy, Development and Environmental Protection (2012). *Prioritetni projekti u oblasti energetike*, (Priority projects in the energy sector, In Serbian).

[29] ICPDR (2013a). *Assessment Report on Hydropower Generation in the Danube Basin*, Viena: ICPDR–International Commission for the Protection of the Danube River.

[30] Yüksel, I. (2009). Dams and Hydropower for Sustainable Development. *Energy Sources, Part B: Economics, Planning, and Policy*, *4*(1), 100-110.

[31] McCartney, M. (2009). Living with dams: managing the environmental impacts. *Water Policy*, *11*(1), 121-139.

[32] Richter, B. D. & Thomas, G. A. (2007). Restoring environmental flows by modifying dam operations. *Ecology and society*, *12*(1), 12 (online).

[33] Kumar, D., & Katoch, S. S. (2014). Sustainability indicators for run of the river (RoR) hydropower projects in hydro rich regions of India. *Renewable and Sustainable Energy Reviews*, *35*, 101-108.

[34] Lenzen, M. (2010). Current state of development of electricity-generating technologies: a literature review. *Energies, 3*, 462–591.

[35] Poff, N. L., Allan, J. D., Bain, M. B., Karr, J. R., Prestegaard, K. L., Richter, B. D., & Stromberg, J. C. (1997). The natural flow regime. *BioScience*, 769-784.

[36] Dudgeon, D., Arthington, A. H., Gessner, M. O., Kawabata, Z. I., Knowler, D. J., Lévêque, C., Naiman, R. J., Prieur-Richard, A. H., Soto, D., Stiassny, M. L. J. & Sullivan, C. A. (2006). Freshwater biodiversity: importance, threats, status and conservation challenges. *Biological Reviews*, *81*, 163-182.

[37] Đorđević, B. & Dašić, T. (2007). Garantovani protoci nizvodno od hidroelektrana (Ecologically guarantied discharge downstream from the hydropower plants, In Serbian). *Elektroprivreda,* 1(januar-mart), 3-11.

[38] Baxter, R. M. (1977). Environmental Effects of Dams and Impoundments. *Annual Review of Ecology and Systematics*, *8*, 255-283.

[39] Liu, J., Chen, F., Cui, Q. & Jiang, Y. (2011). Ecological effect caused by hydraulic engineering construction. *Frontiers of Earth Science*, *5*(2), 170-177.

[40] McCartney, M. P., Sullivan, C. & Acreman, M. C. (2001). *Ecosystem Impacts of Large Dams*. Background Paper Nr. 2, IUCN/UNEP /WCD. http://intranet.iucn.org (accessed June 2014).

[41] Poff, N. L. & Zimmerman, J. K. (2010). Ecological responses to altered flow regimes: a literature review to inform the science and management of environmental flows. *Freshwater Biology*, *55*(1), 194-205.

[42] Rosenberg, D. M., Berkes, F., Bodaly, R. A., Hecky, R. E., Kelly, C. A. & Rudd, J. W. (1997). Large-scale impacts of hydroelectric development. *Environmental Reviews*, *5*(1), 27-54.

[43] World Commission on Dams (2003). *Dams and development–A new framework for decision–making,* London: Earthscan Publications. http://www.internationalrivers.org (accessed June 2014).

[44] Bonacci, O. & Oskoruš, D. (2010). The changes in the lower Drava River water level, discharge and suspended sediment regime. *Environmental Earth Sciences*, *59*(8), 1661-1670.

[45] Magilligan, F. J. & Nislow, K. H. (2005). Changes in hydrologic regime by dams. *Geomorphology*, *71*(1–2), 61-78.

[46] Petković, S., Babić-Mladenović, M. & Damnjanović, M. (2005). Technical, hydrological and environmental aspects of the Iron Gate I and II Dams In: Bloesch, J., Gutknecht, D., & Iordache V. (eds). *Hydrology and Limnology–Another Boundary in the Danube River Basin A*

contribution to the IHP Ecohydrology Project (46-48). IHP-VI Technical Document in Hydrology N°75, Paris: UNESCO/ International Hydrological Programme.

[47] Stefanović, N. & Kapor, R. (2003). Oscilacije vode u akumulacijama prouzrokovane promenama uzvodnog i nizvodnog graničnog proticaja, (Mass oscillations of reservoirs caused by upstream and downstream boundary flow conditions, In Serbian). *Vodoprivreda*, *35*(3-4), 177-187.

[48] Meile, T., Boillat, J. L. & Schleiss, A. J. (2011). Hydropeaking indicators for characterization of the Upper-Rhone River in Switzerland. *Aquatic Sciences*, *73*(1), 171-182.

[49] Radosavljević, P. & Nikić, Z. (2012). Sistemi zaštite od visokih nivoa podzemnih voda priobalja Dunava u zoni akumulacije HE Đerdap (Protection system of high groundwater level in the coastal area of the Danube im the Đerdap I HPP reservoir; In Serbian). In Đukić, A. (ed) *Voda 2012* (Water 2012), Conference Proceedings of the 41th Annual Conference of the Serbian Water Pollution Control Society), 29-36. Belgrade: Serbian water pollution control society.

[50] Babić Mladenović, M. (2007). *Režim nanosa Dunava* (Sediment regime of the Danube River; In Serbian). Beograd: Zadužbina Andrejević.

[51] Babić Mladenović, M., Kolarov, V., Damjanović, V. & Radosavljević, P. (2011).Comparison of the "Natural" and the Present Sediment Regime of the Danube River. *Water Research and Management*, *1*(3), 25-36.

[52] Finger, D., Schmid, M. & Wüest, A. (2006). Effects of upstream hydropower operation on riverine particle transport and turbidity in downstream lakes. *Water Resources Research*, *42*(8).

[53] Klaver, G., van Os, B., Negrel, P. & Petelet-Giraud, E. (2007). Influence of hydropower dams on the composition of the suspended and riverbank sediments in the Danube. *Environmental Pollution*, *148*(3), 718-728.

[54] Salant, N. L., Renshaw, C. E. & Magilligan, F. J. (2006). Short and long-term changes to bed mobility and bed composition under altered sediment regimes. *Geomorphology, 76(1), 43-53.*

[55] *Teodoru, C.* & Wehrli, B. (2005). Retention of sediments and nutrients in the Iron Gate I Reservoir on the Danube River. *Biogeochemistry*, *76*(3), 539-565.

[56] Vörösmarty, C. J., Meybeck, M., Fekete, B., Sharma, K., Green, P. & Syvitski, J. P. M. (2003). Anthropogenic sediment retention: major global impact from registered river impoundments. *Global and Planetary Change*, *39*(1–2), 169–190.

[57] Babić Mladenović, M., Obušković, Z. & Knežević, Z. (2003). Zasipanje akumulacija u Srbiji - problemi i pravci rešavanja (Reservoir sedimentation in Serbia–problems and solutions; In Serbian). *Vodoprivreda*, *35*(5-6), 387-393.

[58] Simović, S., Marković, G. & Milovanović, Z. (2000). Ekološko stanje akumulacije Međuvršje (The ecological status of Međuvršje reservoir, In Serbian). *Vodoprivreda*, *32*(183-185), 220-225.

[59] Sabir, M. A., Shafiq-Ur-Rehman, S., Umar, M., Waseem, A., Farooq, M. & Khan, A. R. (2013). The impact of suspended sediment load on reservoir siltation and energy production: a case study of the Indus River and its tributaries. *Polish journal of environmental Studies*, *22*, 219-255

[60] Ristić, R., Ljujić, M., Despotović, J., Aleksić, V., Radić, B., Nikić, N., Milčanović, V., Malušević, I. & Radonjić, J. (2013). Reservoir sedimentation and hydrological effects of land use changes-case study of the experimental Dičina River Watershed. *Carpathian Journal of Earth and Environmental Sciences*, *8*(1), 91–98.

[61] Hoffmann, T., Thorndycraft, V. R., Brown, A. G., Coulthard, T. J., Damnati, B., Kale, V. S., ... & Walling, D. E. (2010). Human impact on fluvial regimes and sediment flux during the Holocene: Review and future research agenda. *Global and Planetary Change*, *72*(3), 87-98.

[62] Babić Mladenović, M., Radovanović, M. & Radosavljević, P. (2013). Monitoring of the Iron Gate Hydropower and Navigation System on the Danube River. *Water Research and Management*, *3*(1), 3-18.

[63] Brandt, S. A. (2000). Classification of geomorphological effects downstream of dams. *Catena*, *40*(4), 375-401.

[64] Csiki, S. & Rhoads B. (2010). Hydraulic and geomorphological effects of run-of-river dams. *Progress in physical geography*, *34*(6), 755-780.

[65] Graf, W. L. (2006). Downstream hydrologic and geomorphic effects of large dams on American rivers. *Geomorphology*, *79*(3), 336-360.

[66] Babić Mladenović, Kolarov, V. (2010). Hydromorphology In: Simonović, P., Simić, V., Simić, S. & Paunović, M. (eds). *The Danube in Serbia–Results of national program of the Second Joint Danube Survey 2007* (97-110). Belgrade: Ministry of agriculture, forestry and water management–Republic Directorate of water; University of Belgrade, Institute for Biological research Siniša Stanković.

[67] Friedl, G. & Wüest, A. (2002). Disrupting biogeochemical cycles - Consequences of damming. *Aquatic Sciences*, *64*(1), 55-65.

[68] Dević, G., Đorđević, D. & Sakan, S. (2014). Freshwater environmental quality parameters of man-made lakes of Serbia. *Environmental Monitoring and Assessment, 186,* 5221-5234.
[69] Milenković, N., Damjanović, M. & Ristić, M. (2005). Study of Heavy Metal Pollution in Sediments from the Iron Gate (Danube River), Serbia and Montenegro. *Polish Journal of Environmental Studies*, *14*(6), 781-787.
[70] Teodoru, C., McGinnis, D. F., Wüest, A. & Wehrli, B. (2006). Nutrient retention in the Danube's Iron Gate reservoir. *Eos, Transactions American Geophysical Union*, *87*(38), 385-392.
[71] Webb, B. W., Hannah, D. M., Moore, R. D., Brown, L. E. & Nobilis, F. (2008). Recent advances in stream and river temperature research. *Hydrological Processes*, *22*(7), 902-918.
[72] Toffolon, M., Siviglia, A. & Zolezzi, G. (2010). Thermal wave dynamics in rivers affected by hydropeaking. *Water Resources Research*, *46*(8), p.W08536.
[73] Rădoane, M., Ciaglic, V. & Rădoane, N. (2010). Hydropower impact on the ice jam formation on the upper Bistrita River, Romania. *Cold Regions Science and Technology*, *60*(3), 193-204.
[74] Zlatanović, S., Marjanović, P., Marjanović, M. & Milovanović, M. (2013). Oransko zagađenje duž srpskog dela Dunava (Organic pollutin along the Serbian part of the Danube river, In Serbian). *Voda i sanitarna tehnika*, *43*(1), 23-32.
[75] Poff, N. L. & Zimmerman, J. K. (2010). Ecological responses to altered flow regimes: a literature review to inform the science and management of environmental flows. *Freshwater Biology*, *55*(1), 194-205
[76] Renöfält, B., Jansson, R. & Nilsson, C. (2010). Effects of hydropower generation and opportunities for environmental flow management in Swedish riverine ecosystems. *Freshwater Biology*, *55*(1), 49-67.
[77] Bunn, S. E. & Arthington, A. H. (2002). Basic principles and ecological consequences of altered flow regimes for aquatic biodiversity. *Environmental management*, *30*(4), 492-507.
[78] Nilsson, C. & Berggren, K. (2000). Alterations of Riparian Ecosystems Caused by River Regulation Dam operations have caused global-scale ecological changes in riparian ecosystems. How to protect river environments and human needs of rivers remains one of the most important questions of our time. *BioScience*, *50*(9), 783-792.
[79] Gavrilović, Z., Jurišić, S. & Puzović , R (2010). Impact of HEPP "Djerdap I" backwater on foreland forests and revitalization, First

Serbian Forestry Congress - future with forests, 11-13 November 2010, Faculty of Forestry University of Belgrade, Belgrade, 856-871. http://congress.sfb.bg.ac.rs/PDF/erosion/rad14e.pdf

[80] García, A., Jorde, K., Habit, E., Caamaño, D. & Parra, O. (2011). Downstream environmental effects of dam operations: changes in habitat quality for native fish species. *River Research and Applications*, *27*(3), 312-327.

[81] Young, P. S., Cech Jr, J. J. & Thompson, L. C. (2011). Hydropower-related pulsed-flow impacts on stream fishes: a brief review, conceptual model, knowledge gaps, and research needs. *Reviews in fish biology and fisheries*, *21*(4), 713-731.

[82] Paunović, M., Simić, V, Jakovčev-Todorović, D. & Stojanović, B. (2005). Results of investigating the macroinvertebrate community of the Danube River on the sector upstream from the Iron Gate (KM 1083-1071). *Archives of Biological Sciencies*, *57*(1), 57-63.

[83] Jesus, T., Formigo, N., Santos, P. & Tavares, G. R. (2004). Impact evaluation of the Vila Vicosa small hydroelectric power plant (Portugal) on the water quality and on the dynamics of the benthic macroinvertebrate communities of the Ardena river. *Limnetica*, *23*(3), 241-256.

[84] Nat, S. T. S. (2007). Hydroelectric power generation and disruption of the natural stream flow: effects on the zoobenthic community. *Studi Trentini di Scienze Naturali ACTA BIOLOGICA*, *83*, 21-26.

[85] ICPDR (2013b). *Guiding Principles on Sustainable Hydropower Development in the Danube Basin*. Viena: ICPDR–International Commission for the Protection of the Danube River.

[86] Jager, H. I. & Smith, B. T. (2008). Sustainable reservoir operation: can we generate hydropower and preserve ecosystem values? *River research and Applications*, *24*(3), 340-352.

[87] Trussart, S., Messier, D., Roquet, V. & Aki, S. (2002). Hydropower projects: a review of most effective mitigation measures. *Energy Policy*, *30*(14), 1251-1259.

[88] Predlog plana upravljanja Nacionalnim parkom Đerdap za period 2011-2020 god. (Management Plan of National park "Đerdap" for period 2011-2020-proposal, In Serbian) (2012). Donji Milanovac: JP Nacionalni park Đerdap. http://www.npdjerdap.org/publikacije/dokumenta_2012/ppu_npdj_2011_2012.pdf (accessed June 2014).

[89] Roscoe, D. W. & Hinch, S. G. (2010). Effectiveness monitoring of fish passage facilities: historical trends, geographic patterns and future directions. *Fish and Fisheries*, *11*(1), 12-33.

[90] Đorđević, B. & Dašić, T (2011). Određivanje potrebnih protoka nizvodno od brana i rečnih vodozahvata (Method for determing the environmental flow downstream of the dams and water intakes, In Serbian).*Vodoprivreda, 43*(252-254), 151-164.

[91] Dašić, T. & Đorđević, B. (2009). Modeliranje ekoloških procesa u akumulacijama hidroenergetskih sistema (Mathematical modelling of ecological processes in reservoirs of hydroenergetic systems, In Serbian). *Elektroprivreda,* 2 (april-jun), 38-51.

[92] Acreman, M. & Dunbar, J. M. (2004). Defining environmental river flow requirements - a review, *Hydrology and Earth System Sciences*, *8*(5), 861-876.

[93] Lal, R. (2001). Soil degradation by erosion. *Land degradation & development*, *12*(6), 519-539.

[94] Kostadinov, S. (2008). *Bujični tokovi i erozija* (Torrential flows and erosion, In Serbian) Belgrade: University of Belgrade, Faculty of Forestry.

[95] Petković, S., Dragović, N. & Marković, S. (1999). Erosion and sedimentation problems in Serbia. *Hydrological Sciences Journal*, *44*(1), 63-77.

[96] Gostner, W., Lucarelli, C., Theiner, D., Kager, A., Premstaller, G. & Schleiss, A. J. (2011). A holistic approach to reduce negative impacts of hydropeaking In: Schleiss & Boes (Eds). *Dams and reservoirs under changing challenges* (857-865). London: Taylor and Francis Group. ,.

[97] Carter, J. & Howe, J. (2006). The water framework directive and the strategic environmental assessment directive: exploring the linkages. *Environmental Impact Assessment Review*, *26*(3), 287-300.

[98] Pinho, P., Maia, R. & Monterroso, A. (2007). The quality of Portuguese Environmental Impact Studies: The case of small hydropower projects. *Environmental Impact Assessment Review*, *27*(3), 189-205.

[99] Stojanović, B., Milovanović, M. (2011). Primena instrumenata zaštite životne sredine u upravljanju vodnim resursima (The use of instruments for environmental protection in water resources management, In Serbian). In: Stojanović, B. (ed), *Analiza pritisaka i procena uticaja na vodne resurse* (184-207). Beograd: Institut za vodoprivredu Jarosla Černi.

[100] Poff, N. L., Richter, B. D., Arthington, A. H., Bunn, S. E., Naiman, R. J., Kendy, E., ... & Warner, A. (2010). The ecological limits of hydrologic alteration (ELOHA): a new framework for developing regional environmental flow standards. *Freshwater Biology*, *55*(1), 147-170.

[101] Arsić, M., Milivojević, V., Vučković, D., Stojanović, Z. & Vukosavić, D. (2009): Modeling of Flow in River and Storage with Hydropower Plant, Including The Example of Practical Application in River Drina Basin. *Journal of the Serbian Society for Computational Mechanics*, *3*(1), 127-153.

[102] Weller, P. (2013). Reconciling the Interests of Hydropower and Environmental Protection in the Danube River Basin. *Water Research and Management*, *3*(2), 3-8.

In: Hydroelectricity and Power Electronics
Editor: Lucille Burke
ISBN: 978-1-63321-728-7

Chapter 2

THE USE OF HYDROPOWER PLANTS FOR ELECTRICITY GENERATION IN THE EUROPEAN REGION

Jorge Morales Pedraza*
Independent Consultant on International Affairs

ABSTRACT

Hydropower is the major renewable electricity generation technology in Europe today. It delivers storage capacity and stabilizing services for the European power system, which are crucial for a high security of electricity supply. Undoubtedly, hydropower is a mature and cost-competitive renewable energy source and a key component of the energy mix in several EU countries, and this situation will not change during the coming decades. As a multifunctional technology, it is indispensable to the electricity system, and will be even more important tomorrow, when other fossil fuels used today for the generation of electricity, such as oil and coal, will not be sufficient to satisfy the foreseeable EU energy demand.

Hydropower is expected to continue to play an important role in the world energy balance in the future, particularly in the European region. The highest growth rates, however, are expected in developing or strongly industrializing countries with high and yet unexploited hydropower potential, particularly in Eastern Europe. It is important to

* Corresponding author: email jmorales47@hotmail.com.

highlight that since the 1970s, annual energy production in some of the hydropower plants operating in Western Europe has decreased, in particular in Portugal, Spain and other Southern European countries. This reduction has been attributed to changes in average discharge, but whether this is due to temporary fluctuations or already the consequences of long-term changing climate conditions is not yet known.

Keywords: Hydropower; Europe; European Union; generation of electricity; cost of electricity, investment in hydropower; impact on the environment and the population

GENERAL OVERVIEW

Hydropower dates back more than 2 000 years to when the Greeks used water wheels to grind grain. Over the centuries, it has played an important role in providing mechanical energy and, more recently, electricity, supporting human, economic and social development. According to the Swiss Federal Institute of Technology, hydropower is a profitable method to produce electricity. In 2012, hydropower accounts for approximately 20% of total installed renewable energy capacity in the world, thanks to an increase from 793.8 GW capacity installed in 2006 to 1 065 GW capacity installed in 2012[1]. According to EIA sources, the hydropower installed capacity in the EU in 2006 was 161.2 GW; in 2011, the installed capacity reached 171.6 GW, an increase of 6.4% respect to 2006.

Hydropower is the major renewable generation technology in Europe today, representing 17% of the total electricity generation; in the specific case of the EU, the participation of hydropower plants in the generation of electricity reached 10%, according to Eurostat sources. Installed hydropower capacity in Europe is expected to increase 271 GW by 2020. European countries with the largest participation of hydropower in their energy mix include Norway, Latvia, Austria, Sweden, and Croatia, according to Eurostat sources. Maintaining and, in many cases, upgrading this existing infrastructure, continues to be an important focus throughout Europe.

The installed capacity of hydropower plants has grown steadily in the EU over the past 15 years. In 1995, the installed capacity was 90.8 GW, but in

[1] Not included in this amount is the electricity that could be produced by offshore hydropower plants, such as tidal power plants, wave power plants, sea warmth power plants and osmotic power plants.

2008, it was 102.3 GW; this represents an increase of 12.6% respect to 1995. In 2009, the total installed hydropower capacity increased to 136 GW in the EU-27, an increase of 32.9% respect to 2008, with an additional 62 GW in Norway, Switzerland, Turkey, Croatia and Iceland combined. Thus, hydropower represented in that year around 17% of the total installed capacity in the EU-27 (RESAP, 2011). By 2050, it is expected that European hydropower capacity reaches 310 GW; this will represent an increase of 127% respect to 2009.

Besides the countries belonging to the EU, Norway and Switzerland also have significant hydropower assets. In 2008, Norway's hydropower capacity was 29.7 GW, larger than any single EU Member State's hydropower capacity in that year. In the same year, Switzerland's hydropower capacity was 15.3 GW (Ruska and Kiviluoma, 2011).

Hydropower plants deliver storage capacity and stabilizing services for the power system, which are crucial for a high security of electricity supply. Without a doubt, hydropower is a mature and cost-competitive renewable energy source for the generation of electricity and a key component of the energy mix in several EU countries. It plays a key role in today's electricity mix in other regions of the world as well. As a multifunctional technology, it is indispensable to the electricity system, and will be even more important in the future, particularly when other fossil fuels used today for the generation of electricity, such as oil and coal, will not be sufficient to satisfy the foreseeable increase in the European energy demand.

It is important to know that the importance of an electricity generation technology for the overall energy system depends mainly on its capability to stabilize fluctuations between demand and supply. Hydropower plants already contribute to balancing these differences–a function that will be even more valuable in the future, when the reserves of some of the others energy sources currently used for the generation of electricity are depleted.

When assessing the future of hydropower utilization in Europe, two basic influencing factors must be distinguished: (i) a change in exploitable river flow, mainly induced by climate change, and (ii) a change in the existing hydropower installed capacity in the individual countries.

It is important to know that hydropower production varies from year to year depending on hydrological situation. World production of hydroelectricity has grown steadily by about 2.3% per year on average since 1980 (EC, 2000). Worldwide average growth rates of hydroelectricity generation in the future are estimated from about 2.4% per year between 1990 and 2010 (Voigtländer and Gattinger, 1999) to 3.6% per year between 2010 and 2020 (Eurelectric, 1997). The highest growth rates are expected in developing or strongly industrializing countries with high, but yet unexploited hydropower potential, e.g., parts of Eastern Europe, while in Western Europe only a 1% annual increase is expected (Voigtländer and Gattinger, 1999). In the specific case of the EU-27, in 2008 the total hydropower production was 564.4 TWh, according to EIA sources. Two years later, in 2010, the hydropower production was 615.2 TWh; this represents an increase of 9% respects to 2008. In 2011, the hydropower production in the EU-27 reached 540.1 TWh; this represents a decrease of 12.3%, but with the possibility to increase once again during the coming years.

According to the total installed hydropower capacity in the EU-27 and the level of electricity generated by this type of energy source, hydropower accounts for the lion's share of renewable electricity generation in Europe (70%). It is followed by wind (15%), primary solid biomass (7%), municipal waste (4%), biogas (3%), solar (1%) and geothermal (1%) (RESAP, 2011). An estimated 30 GW of new hydropower capacity came on line in 2012, increasing global installed capacity by 3% to an estimated 1 065 GW. Hydropower generated an estimated 3 700 TWh of electricity during 2012, an increase of 6.5% respect to 2011. Once again, China led in terms of capacity additions (15.5 GW), with the bulk of other installations in Turkey, Brazil, Vietnam, and Russia, the only European country included in this group. Joint-venture business models involving local and international partnerships are becoming increasingly prominent as the size of projects and the capacity of hydropower technologies increase (REN 21, 2013).

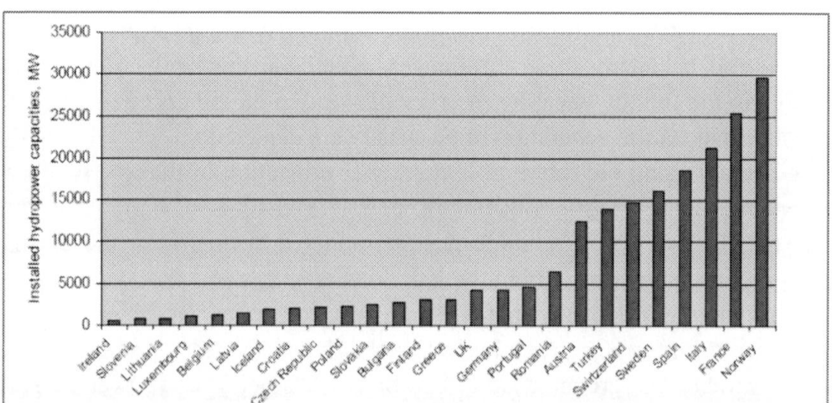

Source: RESAP (2011).

Figure 1. Installed hydropower capacities in MW in Europe in 2009.

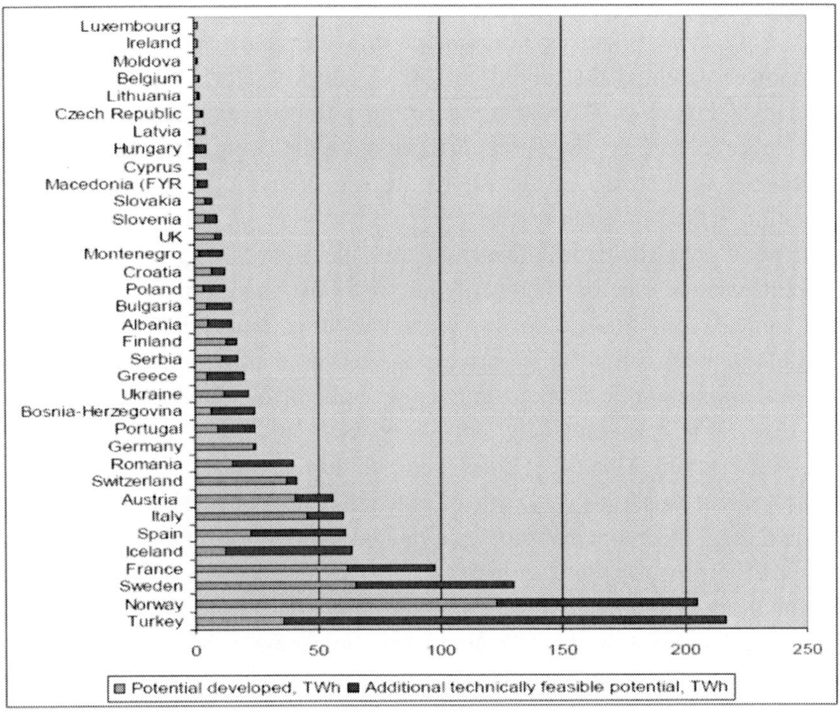

Source: RESAP (2011).

Figure 2. Developed and remaining technically feasible hydropower potential in Europe per country in TWh.

The total installed hydropower capacity in non-EU Member States (Albania, Bosnia-Herzegovina, Former Yugoslav Republic of Macedonia, Moldova, Montenegro, Serbia and Ukraine) is now 12.5 GW (RESAP, 2011).

According to EU sources, hydroelectric is the most important renewable energy source within the European region from a global perspective. Although most of the best sites for hydropower plants have already been developed in Europe, at present only about half of its technically feasible potential has been developed, and only about one third in Albania, Bosnia-Herzegovina, Moldova, Former Yugoslav Republic of Macedonia, Montenegro, Serbia, and Ukraine (World Atlas and Industry Guide 2010).

Figure 2 shows developed and still available technically feasible hydropower potential per country in TWh. It is important to highlight, that Denmark and the Netherlands have a very small hydropower generation today and with comparatively little development potential; Malta has neither hydropower generation, nor any potential. In total, these countries make up only 0.178 TWh of hydropower generation a year and represent 0.6 TWh of hydropower potential that could still be developed. They are, therefore, not included in Figure 2. From that figure the following can be stated: France, Italy, Turkey, Norway, Sweden and Iceland are the countries with the highest possibilities to increase in the future the use of hydropower for electricity generation in the European region.

In addition to the hydrological situation, the role of hydropower within the different national electricity generation portfolios varies greatly according to geographical conditions, climate, precipitation patterns, the availability of affordable energy supply alternatives, as well as institutional capacities and technical competences. Some countries have abundant water resources, which they use almost exclusively for their domestic electricity generation (e.g., Norway and Albania). Other countries do not have the required natural conditions and thus have very little or practically no hydropower production or potential (e.g., Denmark and the Netherlands) (RESAP, 2011).

Finally, it is important to highlight that hydropower has the potential to become more economically attractive via improved turbine designs and cost-effective plant construction in combination with new technologies of control and operation. From an operational standpoint, a hydropower plant can start very quickly and the power output can be controlled accurately and for this reason this type of power plants will be more attractive, from the economic and technical point of view, in the future.

Types of Hydropower Plants

Hydropower plants convert the kinetic energy contained in falling water into electricity. The energy in flowing water is ultimately derived from the sun, and is therefore constantly being renewed. The energy contained in sunlight evaporates water from the oceans and deposits it on land in the form of rain. Differences in land elevation result in rainfall runoff, and allow some of the original solar energy to be captured as hydroelectric power.

According to their scale, hydropower plants can be classified as large and small. Large-scale hydropower dominates the production of electricity with 86.7% of the total electricity generated in 2010, while small-scale hydropower production was only 13.3% of the total. According to the European Renewable Energy Council, small hydropower is based on a simple process, taking advantage of the kinetic energy and pressure freed by falling water or rivers, canals, streams and water networks. The rushing water drives a turbine, which converts the water pressure and motion into mechanical energy, converted into electricity by a generator. The power of the scheme is proportional to the head (the difference between up and down stream water levels), the discharge (the quantity of water, which goes through the turbines in a given unit of time), and the efficiency of the turbine.

Small hydropower plants currently contributes more than 37 TWh per annum or 2.5% of the European electricity market. Modernization, reconditioning and the exploitation of new sites will mean that around 50% of the remaining small hydropower sites in Europe will be exploited by 2015. Undoubtedly, small hydropower plants is one of the most widely used renewable resources in the world for the generation of clean power, but the global hydropower market is expected to continue its growth path with a major contribution from large hydropower plants during the coming decades.

Besides the countries belonging to the EU, Norway and Switzerland also have significant electricity generation using hydropower plants. During the coming years, largest additions in large-scale hydropower production are expected in Spain (4.5 TWh), Portugal (3.6 TWh), Austria (3.1 TWh), and Romania (2.6 TWh). Two countries, Sweden and Italy, expect large-scale hydropower production to decrease by 3.2 TWh (Sweden) and 3 TWh (Italy), respectively. The largest small-scale hydropower addition within the EU-27 is expected in Italy (2.9 TWh).

Hydropower plants can also be classified according to the way they make use of the inflow that they receive, depending on whether this cumulative flow must be used within a short period of time or whether it can be retained for a certain time period (UCTE, 2000). The classification of the current hydropower plants distinguishes between run-of-river, pondage and reservoir, and pumped hydropower plants.

Hydropower plants can also be divided, based on the operational mode, into run-of-river and reservoir (pondage, reservoir and pumped) hydropower plants. Run-of-river hydropower plants use the flow of water from upstream without any substantial storage. This may be a natural flow or flow determined by other water use upstream. Run-of-river hydropower plants use the flow of water within a river's natural range. This type of hydropower plant shows fluctuations in energy production induced by low flow or overflow periods, whereas reservoir hydropower plants can store water (i.e., energy) over long time periods and generate a steady supply of electricity, relatively independent of variations in short-term inflows (Lehner et al., 1998).

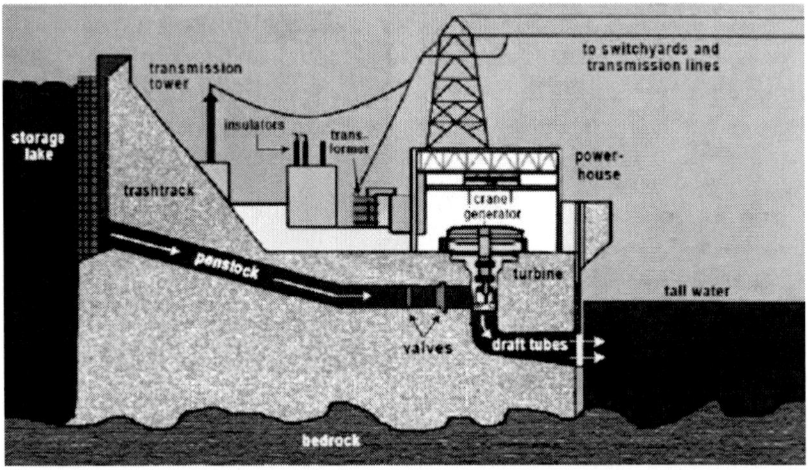

Source: IEA.

Figure 3. Principle of a hydropower plant.

Source: Verbund, Austria.

Figure 4. Freudenau run-of-river hydropower plant in Austria (a low head run-of-river hydropower plant).

The electricity output of a run-of-river hydropower plant depends on the availability of water in the river and varies considerably throughout the year. Seasons, rainfall or snowmelt all influences the amount of water available for electricity generation. Because they lack any kind of storage, run-of-river hydropower plants operate under the constraint of precisely controlling the water level at the intake in accordance with the incoming river flow. Run-of-river hydropower plants without connection to upstream storage typically operate as base load power plants since the hydrological forecast is sufficiently good for the timescales required in the electricity market (RESAP, 2011).

Upstream hydropower plants or reservoirs have to be taken into consideration when looking at the energy output of a run-of-river hydropower plant. A constant output from this type of power plant is possible, if the river is already regulated further upstream by a hydropower plant with storage capacity (thanks to the reservoir). A commonly used strategy to optimize the energy output of hydropower plants on a river is to build a large storage reservoir in the upper catchment, which will even out flows for several run-of-river or smaller reservoir hydropower plants downstream. Beyond the energy storage function of reservoir hydropower plants, regulating a river in the upper area of a river basin with a reservoir will increase the energy potential of sites downstream, as the regulated river will typically flow more evenly throughout the year. Multiple run-of-river hydropower plants can be developed downstream of a reservoir type hydropower plant. That way, water is re-used to produce additional electricity as it flows down the same river. The combined cascade of dams and reservoirs allows an optimized electricity generation and may also be used to absorb excess energy when reducing river flow, thereby strongly enhancing the storage function from the upper reservoir.

Pumped storage hydropower plants use excess electrical system capacity, generally available at night, to pump water from one reservoir to another reservoir at a higher elevation. During periods of peak electrical demand, water from the higher reservoir is released through turbines to the lower reservoir, and electricity is produced. Their value is in their ability to store electricity for use later when peak demands are occurring, but this type of hydropower plants are even more valuable, if intermittent sources of electricity, such as solar or wind power sites, are hooked into a system.

With regards to pumped storage hydropower plants, it is important to highlight the following: In the EU, electricity generation from pumped storage hydropower plants is not considered as renewable, since pumping the water up consumes electricity produced by other power plants in the grid and pumped storage hydropower plants thus become net energy consumers. Overall efficiencies of this type of hydropower plant are in the range of 70% - 75%. However, this disadvantage is compensated by the flexibility that this type of plants can provide to the system.

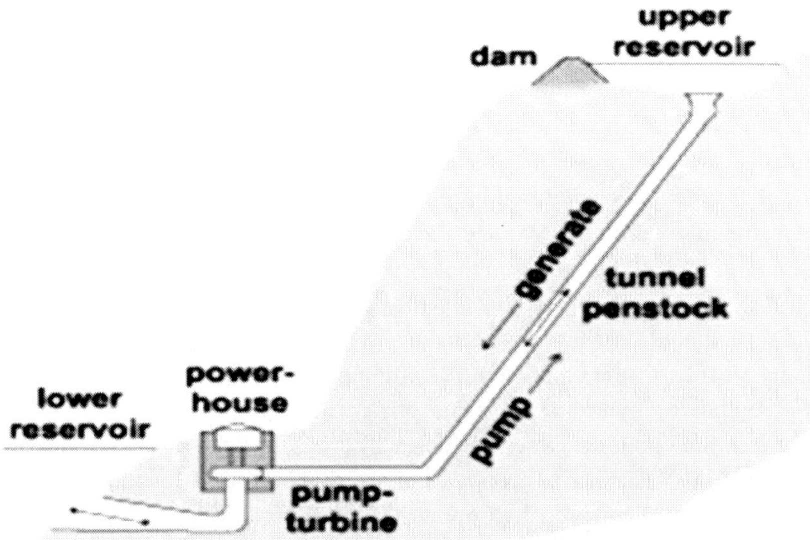

Source: Mechanical Writings.

Figure 5. Illustrations of the operating principle of a pumped storage hydropower plant.

Pumped storage hydropower plants are important to optimize the operation of the conventional thermal generation power plants connected to the grid. As pumped storage hydropower plants cover peak demand situations while consuming electricity in periods of lower demand, they allow thermal power plants to stay connected and generate electricity even when demand is low. Through this optimization effect, pumped storage hydropower plants also contribute to reducing greenhouse gas emissions from thermal power plants, as they need not deviate from the most efficient load. The role of pumped storage hydropower plants is twofold: balancing the grid for demand-driven fluctuations, and balancing generation-driven fluctuations–a new and increasingly important role due to the increase of the participation of renewables in the electricity system (RESAP, 2011).

Finally, it is important to highlight the following: Like a conventional reservoir-type hydropower plants, pumped storage hydropower plants can provide the full range of grid-stabilizing services, thanks to their ability to follow demand or generation fluctuations within only a few minutes. Therefore, a storage hydropower plant is a key tool for TSOs to maintain a stable and balanced grid.

The evolution of the construction of pumped storage hydropower plants during the period 1990-2015 is shown in Figure 6.

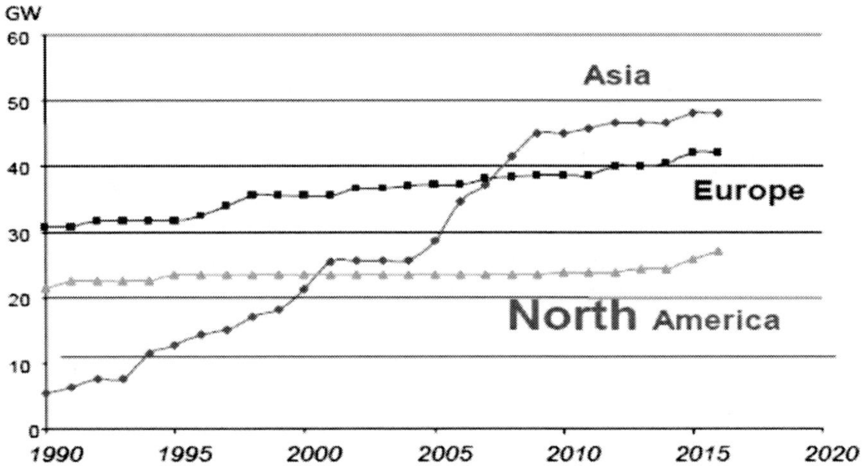

Source: EDF R&D.

Figure 6. Evolution of the construction of pumped storage hydropower plants during the period 1990-2020.

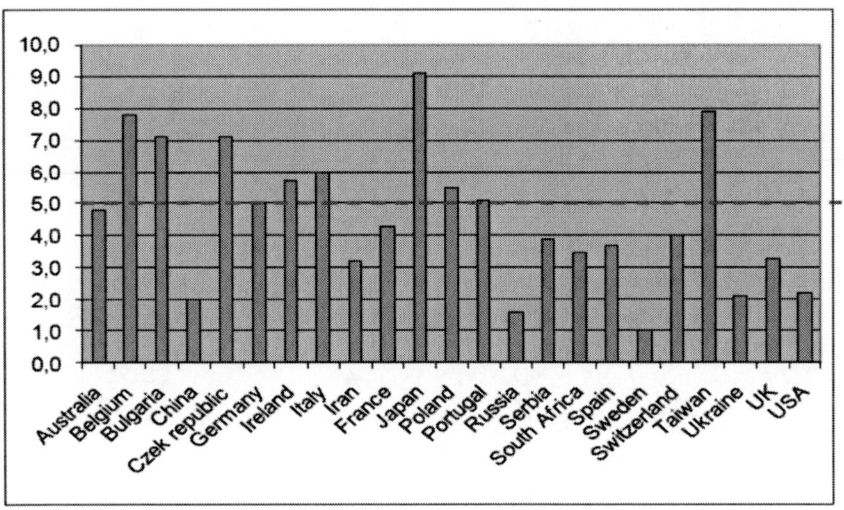

Source EDF.

Figure 7. Percentage of pumped storage hydropower plants per installed capacity by country.

From Figure 6, the following can be stated: Asia is the region with the highest increase in the construction of pumped storage hydropower plants during the period 1990-2010, and it is expected that this trend will continue without change between 2010 and 2015. Europe was the region with the highest increase in the construction of this type of power plants until 2007, and falls to the second position after that year.

The percentage of pumped storage hydropower plants per installed capacity is shown in Figure 7. According to that figure, the countries with the highest percentage of pumped storage hydropower plants per installed capacity are Japan, Taiwan, Belgium, the Czech Republic, Bulgaria, Italy, Ireland and Poland. In this group, 6 countries are located within the European region (75% of the total).

The main characteristics of the pumped storage hydropower plants are, among others, the following:

- It is a mature and flexible technology in use in several countries in all regions;
- It is a fast acting power plant;
- The performance of this type of hydropower plants is very high;
- Provide electricity when needed;
- It is an effective technology for load balancing;
- Has a limited impact on the environment;
- Can be in exploitation between 40 and 80 years, but with good maintenance could be in exploitation up to 100 years.

THE LIFECYCLE OF HYDROPOWER PLANTS

The lifecycle of hydropower plants consists of three main stages. These are the following:

- **Construction:** This stage includes production and transportation of materials (e.g., concrete, steel, etc.) and the use of civil work equipment and materials for the construction of the facility. This is the stage with the highest consumption of energy and has the greatest impact in the environment and the population;
- **Operation and maintenance:** This stage includes all operation and maintenance activities related to the exploitation of the hydropower plant. According to different sources, the operation of a hydropower plant could be between 40 and 100 years, if good maintenance is

given to the plant. The average cost of operation and maintenance of a hydropower plant has been accepted to be around 2.5% of the total cost;

- **Dismantling:** This stage includes the decommissioning of the plant for economic, safety or environmental reasons.

THE USE OF HYDROPOWER PLANTS FOR ELECTRICITY GENERATION IN THE EUROPEAN REGION

The role of hydropower, along with other renewable energy sources for the generation of electricity within the European region, is expected to become increasingly important in the future[2]. Why? First, because, in Europe around 36% of the economically viable hydropower potential is not being tapped, despite of the fact that hydropower is the dominant renewable electricity source in the region. The potential hydropower theoretically to be tapped in Europe (excluding Russia) totals nearly 2 600 TWh per year. Currently, 64% of the economically viable potential is being exploited. It follows that Europe leaves no less than 36% of its hydropower potential unused, even though related power generation would pay off. In Europe, by the 2070s, hydropower potential for the whole region has been estimated to decline between 20% and 50% around the Mediterranean, between 15% and 30% will increase in Northern and Eastern Europe, and a stable hydropower pattern for Western and Central Europe is expected to occur (Lehner et al., 2005).

Second, it is well known that the electricity system can only function properly, if electricity generation (supply) matches electricity demand (consumption). Planning generation, according to the forecasted demand is therefore essential because electricity demand is variable throughout the year and it depends on the period of the year under consideration. For example, more electricity is consumed in winter than in summer. However, daily electricity demand always follows a similar pattern. These plans, however, are not always met. Therefore, any change in electricity demand or generation needs to be met by an equivalent adjustment either on the supply or the

[2] Hydropower plants may, depending on their design, provide electricity for base load and/or peak load. They are particularly valuable for meeting peak demand situations, as they are more responsive than other generation sources and can be started or stopped within a very short time. Hydropower units are also able to rapidly increase or decrease their output – at least 10 times faster than conventional power plants.

demand side. In other words, there is a need for balancing the plans. This is done through market actions or in some cases via TSOs.

In addition to these well-known variations in demand patterns, an increasing amount of renewable energy sources contributes significantly to the challenge of keeping the electricity system in constant balance. In a purely conventional power plant portfolio the load or the required/delivered amount of electric power falls into three categories: base load, intermediate load, and peak load[3]. Base load refers to a relatively constant output of power plants over a period. It is typically delivered by power plants that are designed to generate electricity at a constant rate of output. In contrast, peak load refers to surges in electricity demand that occur at specific, usually predictable periods (i.e., evening peak load, when consumers simultaneously switch on lighting and other electric appliances). Due to their greater responsiveness, peaking power plants are able to provide electricity when demand is at a high. Finally, intermediate load refers to a situation where power output increases every morning and throughout the day and decreases in the evening. Intermediate load power plants run considerably more hours than peaking units, but fewer hours than base-load power plants (RESAP, 2011).

On the other hand, hydropower is receiving a strong boost from the rising electricity prices that can be expected as a result of increasing internalization of climate costs, the growth in global energy consumption and the associated growing scarcity of fossil fuel sources. It is expected that the European electricity prices to rise by an average of around 4% per year until 2030. Major developments are likely to be a significant increase in the price of fossil fuels, rising infrastructure expenditure as well as environmental and climate policy readjustments (Auer, 2013).

With the scene set for an increasing share of renewable energy sources in the energy mix of EU Member States, storage hydropower plants will become more important. Storage hydropower plants can bridge the gap between demand and supply and support the integration of renewable energy sources into the electricity system. Hydropower storage plants are the only large-scalc storage technology available today. They are also the most efficient and

[3] Different types of electricity-generating plants are differently suited to operation as base-load plants or peaking units. In general, nuclear power plants, run-off-river hydropower plants and conventional thermal power plants are covering the base load demand. Utilities like gas-fired power plants and storage hydropower plants cover the peak load demand. Pump storage facilities are the only renewable energy source, which can support the system by highly flexible power generation ancillary services and peak load generation. Pump storage hydropower plants are enabled to replace renewable or thermal power plants from base load to peak load generation (RESAP, 2011).

economical way to store potential electricity. While pumped storage hydropower plants have a short to medium-term storage capacity depending on the size of its reservoirs, conventional storage hydropower plants can offer significant long-term storage capacity. Although pumped storage hydropower plants are a net electricity consumer, it provides valuable regulatory services to stabilize the electric system. Moreover, it can be used to store excess production of renewable energy sources. Compared to all other storage technologies, pumped storage hydropower plants represent the most cost efficient option[4] (RESAP, 2011).

Besides the fact that hydropower currently makes up a substantial share of the total amount of electricity generated in the world, the arguments for the continued and increased utilization of hydropower plants for the generation of electricity are based on its advantages when compared to other sources of energy. These are:

- Low-cost[5];
- Effectiveness;
- Sustainability;
- Stored in large quantities[6] (Lehner et al., 1998).

[4] Multipurpose hydropower projects could also provide services beyond the electricity sector. They contribute to improved water management, such as flood and drought control, which could prove especially valuable in the context of climate change adaptation. As for the decarbonization agenda, hydropower has been identified as highly valuable for climate change mitigation, due to its low carbon footprint and high generation efficiency. This makes hydropower the most competitive and reliable renewable energy source (RESAP, 2011).

[5] According to IEA, the cost of power production from hydropower can vary widely depending on project details, but usually fall into a range of US$ 50 to US$ 100 per MWh.

[6] Storage hydropower plants can often be operated flexibly, and therefore are valuable to electric power systems. Specifically, with its rapid response load-following and balancing capabilities, peaking capacity and power quality attributes, hydropower can play an important role in ensuring reliable electricity service. In an integrated system, reservoir and pumped storage hydropower plants can be used to reduce the frequency of start-ups and shutdowns of thermal power plants; to maintain a balance between supply and demand under changing demand or supply patterns and thereby reduce the load-following burden of thermal power plants; and to increase the amount of time that thermal units are operated at their maximum thermal efficiency, thereby reducing carbon emissions. In addition, storage and pumped storage hydropower plants can help reduce the challenges of integrating variable renewable resources such as wind, solar photovoltaic, and wave power (Kumar et al., 2011).

Electricity generation from large hydropower plants is characterized by:

- **High exploitation and proven technology**: Among all renewable energy sources used for electricity generation, hydropower represents the most explored sources;
- **Low volatility of the power output**: Hydropower represents a fluctuating source of energy. In contrary to wind and solar photovoltaic, the volatility appears in the medium to long-term. It is characterized by a seasonal dependence, but also high annual differences occur;
- **Strong public opposition**: Public resistance to the construction of new large-scale hydropower plants has been raised in most parts of Europe since the 1980s when new large-scale projects have been discussed. For this reason, the environment and social effects of hydropower projects need to be carefully studied before any decision to construct new large-scale power plants is taken within the region.
- **High initial investment costs**: One of the major issues associated with the construction of large-scale hydropower plants is the high initial investment costs associated with the construction of this type of power plants;

Countries should follow an integrated approach in managing their water resources, planning hydropower development in co-operation with other water-using sectors, and take a full life-cycle approach to the assessment of the benefits and impacts on any new construction of large-scale hydropower plants under consideration. Over the last two decades, decisions on many large-scale hydropower projects have been affected by the controversy about the environmental and social effects of this type of renewable energy sources for the generation of electricity. In deciding on the role of hydropower within an electricity supply portfolio, which addresses climate change and energy security concerns, energy policy makers have to consider a whole range of issues, including:

- Enhancing economic equity among citizens;
- Protecting the lives and property of citizens from floods and droughts;
- Securing the rights of citizens with respect to the expropriation of land to be inundated; and
- Protecting the environment concerning air, land, water and biodiversity;

Investment in Hydropower Plants

Investment in large hydropower plants of more than 50 MW continued to be significant in 2012, exceeding all other renewable energy sectors, except wind and solar power. Translating capacity additions into asset finance dollars per year is not straightforward because the average hydropower plants take four years to build, but it is estimated that asset financing for hydropower projects commissioned in 2012 totaled at least US$ 33 billion—over a fifth of the US$ 148.5 billion value of asset finance, excluding large hydropower plants.

Investment opportunities lie in countries with huge potential in the Alps and in Scandinavia, where the use of hydropower plants for the generation of electricity already has a long tradition. The convergence of regional European electricity markets as well as technical megaprojects like the offshore North Sea Ring, will make hydropower plants even more interesting in the future within the European region.

Southeastern Europe utilizes just 40% of its hydropower potential. Since 60% of the economically viable hydropower potential is still waiting for investors, it is worth also taking a look at Southeastern Europe. Specifically because the region's electricity supply largely relied on centrally planned large-scale structures for more than half a century of being constructed, there are now lots of interesting project opportunities that elsewhere in Europe that can only be found in isolated cases (Auer, 2013).

Several new hydropower projects entered into commercial operation in the past few years, something not seen in several decades. Examples of new projects include: Sonna in Norway (270 MW), Glendoe in the UK (100 MW), and Blanca in Slovenia (42.5 MW).

For small hydropower plants, development opportunities are significant. Provided the mandate by EU Member States is implemented on a timely basis, the European Small Hydropower Association estimates that installed small hydropower capacity could reach 16 GW by 2020–a more than 4 000 MW increase over current levels.

Another area of significant growth for the hydropower sector in Europe, especially in the central region of the continent, is in pumped storage facilities. In addition to supplying additional electricity during times when demand for power is high, pumped storage's ability to balance power production and regulate the transmission network, in light of increased use of intermittent renewable, particularly wind and solar, is attractive. As many as 10 pumped-storage facilities are under construction in the European region, including 178

MW Avce in Slovenia, 540 MW Kopswerk 2 and 480 MW Limberg 2 in Austria, and 141 MW Nestil in Switzerland. Several more potential hydropower projects are being considered.

The emphasis in Western Europe is not in the construction of new large-scale hydropower plants due to a strong public opposition, but in retrofitting hydropower plants with modern equipment, usually upgrading the capacity of the plant. In Eastern Europe, the focus is rehabilitating aging hydropower plants constructed after 1945. Numerous utilities are committing significant resources to upgrade several hydropower plants located in different European countries. For example, in France, national utility Electricite de France is investing more than US$ 2.5 billion as part of France's economic stimulus program, including spending on modernization of hydroelectric projects.

Finally, it is important to highlight the following: Energy prices and the political measures towards CO_2 reduction in the whole EU will have a positive impact on the level of investment allocated to the hydropower sector. Decisions on power plant projects are directly influenced by expectations concerning the price of energy inputs, the price elasticity of energy demand, and the availability of competitive domestic sources of energy. Measures for decarbonize industry may indirectly increase the cost of using fossil fuels for electricity generation and thus raise questions about all the related calculations. Moreover, the prices of procuring and transporting the energy inputs are by no means constant, since (unexpected) fuel price increments may also jeopardize the profitability of investments (Auer, 2013).

Benefits and Relevance of the Use of Hydropower Plants for the Generation of Electricity

According to the European Renewable Energy Council, the main benefits and relevance to the use of hydropower plants for the generation of electricity are obvious. In particular, the use of hydropower plants is expected to play a key role in 2030 and 2050 within the European region though:

- Development of hybrid systems combining several technologies to guarantee the maximum energy production in the most efficient way;
- Development of multipurpose hydropower plants with applications in the fields of drinking water supply systems, irrigation channels, flood

control and protection, the creation of adjoining environmental areas, waste water treatment plants, and recreational purposes;

- Adding security and stability to the European grid thanks to the use of pump storage hydropower plants;
- Mitigating climate change;
- Since it does not require any fossil-based primary energy, the use of hydropower plants for the generation of electricity provides relief for the energy bill of every country and increases the security of the energy supply;
- Supporting the development of rural areas by the installation on economically advantageous conditions of small hydropower plants off-grid units;
- Hydropower has, according to Auer (2013), further advantages over many of the new renewable: one example is efficiency levels of up to 85% taking all types of hydropower plants into consideration. This makes hydropower a more efficient type of electricity generation than solar cells or wind farms, which achieve efficiency levels of 20% and 40%, respectively, in practice. In addition, hydropower is in most cases independent of the time of day, i.e., is not governed by the amount of sunshine or wind speed;
- All the same, investments in the construction of hydropower plants are marked by relatively high initial outlays that are subsequently followed by very low operating costs (less than 1 cent per kWh). This is a major advantage over the use of fossil-fired power plants for the generation of electricity, where the risk of increasing input costs for coal, natural gas and oil can jeopardize returns;
- State-of-the-art hydropower plants can pave the way to multiple-use facilities. Improvements in flood protection, shipping conditions and the watercourse downstream from hydropower plants may be integrated. Moreover, agricultural and recreational areas may also benefit, as their attractiveness increases.

The Future of the Hydropower Sector in the European Region

Analysis of Europe's power plants reveals two major trends. First, the electricity industry is banking less and less on coal, nuclear power and oil in its investment decisions in several countries within the region. For each of these power generating alternatives, more capacities were removed from the grid than new ones were installed. The net change in capacities installed in the EU totaled minus 12 GW in the case of coal-fired power plant, minus 7.2 GW for nuclear power plants and minus 12.9 GW for oil-based generation power plants during the period 2000-2009. Second, there is a strong trend visible in the power generation sector towards the increase in the use of renewable energy sources of electricity generation, and thus towards hydropower, as well as natural gas. The net growth in the case of natural gas totaled 81 GW, wind power 65 GW, solar photovoltaic 13 GW and hydropower 3 GW (Auer, 2013).

It is important to highlight that hydropower is expected to continue to play an important role in the world energy balance in the future. By 2020, a fifth of all energy consumption in EU Member States must come from renewable energy sources. This mandate, which EU leaders signed in March 2007, is part of a proposal designed to cut greenhouse gas emissions by 20% (compared with 1990 levels). However, it should be noted that in the specific case of hydropower projects, they have also negative effects concerning environmental and social issues. This negative effect must be considered when planning the construction of new hydropower plants, particularly within the EU.

The share of hydropower within the total electricity generation varies considerably between countries, ranging from negligible to 99%. These differences in countries reflect both geographic and climatic constraints and suitability, as well as, to some degree, political guidelines in operation (Lehner et al., 1998).

In Eastern Europe most of the development of hydroelectricity in the short to medium-term is expected to occur as expansion or refurbishment of existing hydropower plants, as the persistent economic problems interfere with the construction of new plants. In East-Central Europe, hydroelectricity already represents a substantial source of power in some countries such as Albania (96 %), Croatia (59%) or Romania (37%) of total electricity generation. Most of the potential for future hydropower expansion lies in Albania, Bulgaria and Romania, as well as in the former Yugoslav republics. However, despite a very large potential for future expansion, yet, these countries have found it difficult to secure financing for such projects (EC, 2000).

The Nordic European countries generally show good suitability for hydroelectricity production, both from geographic and climatic aspects, and already have high shares of hydropower forming part of their energy supply. However, in the case of Sweden, for example, there is no scheme for building new large-scale hydropower plants, as proposals for this meet strong public opposition. Only old units are refurbished and some new small units are installed. Consequently, no major additions to the current hydropower production capacity are expected. Also from Norway, only small increases of the cumulative volume of hydropower plants are expected (Lehner et al., 1998).

Latvia, Portugal, Slovenia, Spain, France and Italy are the European countries where current hydropower production represents a significant share of the total renewable energy contribution.

From Figure 8, the following can be stated: The immense capacity in hydropower is associated with plants with a capacity higher than 10 MW, followed by hydropower plants with a capacity higher than 1 MW.

The following are the European countries with the highest number of hydropower plants (with a capacity >10 MW) used for the generation of electricity: Norway (around 27 500 MW), France (around 22 000 MW), Sweden (a little bit higher than 15 000 MW) and Italy, Switzerland and Spain (around 14 000 MW).

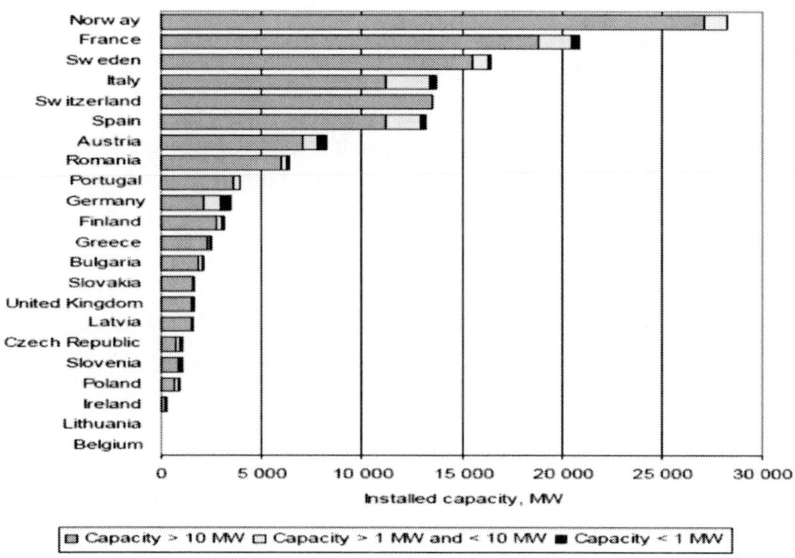

Source: Ruska and Kiviluoma (2011).

Figure 8. Electricity generation from hydropower plants in the EU.

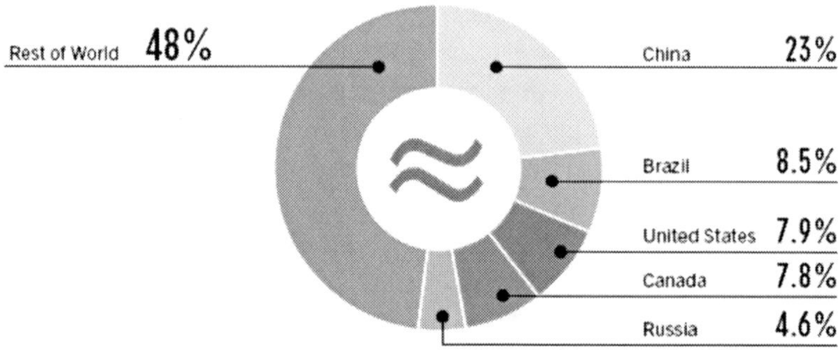

Source: REN 21 (2013).

Figure 9. Hydropower installed capacity, shares of top five countries (2012).

According to Figure 9, Russia is the only European country included in the top five countries with the highest percentage in hydropower capacity installed in 2012 (4.6%).

Finally, in order to achieve the forecasted vision of the sector for 2030 and 2050 or surpass these estimations, the following issues must be tackled:

- **Reconciling targets of the Water Framework Directive (WFD) and the RES Directive:** The implementation of the WFD is currently restraining the present and future development of the hydropower sector. The reason is the different interpretation of the Directive at a national level, and this situation is having direct consequences in terms of the approval of new hydropower projects and in terms of the allocation of concessions and permits in the different EU Member States;
- **Environmental measures:** Hydropower needs a more objective approach of the environmental community and from stakeholders for current and future legislation because it could limit in a severe way the benefits of the use of this source of energy for the generation of electricity in the different EU Member States;
- **Removal of administrative and regulatory barriers:** Administrative procedures to get a hydropower plant operating are still one of the most important barriers for an increase role of the hydropower sector within EU Member States. Longer time periods required for obtaining licenses, concessions and permits for the construction of new hydropower plants discourage developers from

bringing projects to an end. A more flexible, simple, centralized and homogeneous European system could ease the procedure;

- **More attractive incentive regimes (especially in the new EU Member States):** Hydropower and in particular small units are currently benefiting from European support schemes. Nevertheless, in comparison to other renewable and comparing between countries, the level of support is not satisfactory in terms of cost-benefit and market competition;
- **Need for proactive cooperation and better communication at a local level:** In the case of hydropower projects, the rapid establishment of a participatory approach involving the different stakeholders affected by the realization of the project, and in particular the environmental and fishing community, is a must for the future development of the hydropower sector;
- **Investment in research and development and change of thinking:** The hydropower technology of the next decades will evolve towards more sustainable solutions. However, in order to minimize the environmental impact while at the same time maximizing electricity production, a change of thinking is required and investment in current and future research and development is highly recommended to explore and test different solutions.

Though hydropower is a proven and well-advanced technology, there is still room for further improvement, for example, through optimization of operation, mitigating or reducing environmental impacts, adapting to new social and environmental requirements, and more robust and cost effective technological solutions, just to mention a few ideas for further consideration. Large hydropower turbines are now close to the theoretical limit for efficiency, with up to 96% when operated at the best efficiency point, but this is not always possible and for this reason continued research is needed to make more efficient operation possible over a broader range of flows. Most of the existing hydropower equipment in operation today will need to be modernized during the next three decades, allowing for improved efficiency and higher power and energy output (UNWWAP, 2006), but also for improved environmental solutions by utilizing environmental design principles.

Negative Impact on the Environment and on the Population due to the Use of Hydropower Plants for Electricity Generation

The local social and environmental impacts of hydropower projects vary depending on the project type, size and local conditions, and are often controversial. Some of the more prominent impacts include changes in flow regimes and water quality, barriers to fish migration, loss of biological diversity, and population displacement. Impoundments and reservoirs stand out as the source of the most severe concerns, but can also provide multiple beneficial services beyond the energy supply (Kumar et al., 2011).

Renewable energy plants such as biofuels, wind power, solar photovoltaic and hydropower, can be constructed and operated in a way that causes only marginal emissions of carbon dioxide and less pollution of e.g., sulphur and nitrogen oxides. In the case of hydropower, this type of energy source has an important impact in the industrial development of many countries and plays an important role today in providing the electricity they need to satisfy energy demand.

However, run-of-the-river hydropower plants substantially change the river environment. The installations destroy habitats, decrease the number of species and severely alter the river morphology. The type and size of the plant will give different impacts. Many of the large-scale hydropower plants have storage reservoirs, created by damming the river and flooding the area upstream of the generator. The flooded area impacts the people, the vegetation and the animals in the area. The dried flood bed downstream of the dam drastically deteriorates the living environment for all the species living in and around the river, causing river fragmentation and obstacles for fish migration.

Hydropower plant constructions, where water is dammed and a storage reservoir is created, will have additional environmental impacts in connection to the dam. The dam will flood the area upstream of the hydropower plant and the river bed downstream will be dried. The dam will have a severe impact on people living in the area, the natural environment will be destructed and the construction and operation of a dam always involve certain safety risks (Sørensen, 2000). Different stages of a dam construction will put a different type of stress on the environment. The effects can occur as direct consequences of the dam as well as indirect consequences. The construction of the hydropower plant and the creation of a dam have the impact of causing obstacles in the river, destructive elements connected to the construction work, population displacement and modification of the landscape. These direct effects of the dam construction will in turn cause many indirect effects. This includes barriers to migration of aquatic fauna, disruption of habitats and on wildlife in and around the river, increased sediment transport and turbidity in the water. The creation of a dam is significantly changing the landscape, but can, in some cases, lead to increased tourism and recreation as well as a new habitat (Leonard, 2000).

The fragmentation caused by a hydropower plant, as well as other structures in the water, has a severe impact on the habitats and the flora and fauna in the riverine environment. Many of the species living in the rivers are depending on migrations along the water course and to be able to colonize new habitats. The migration of fish is disrupted by the operation of the hydropower plant, but many other species are affected as well. Several aquatic insects spread in the riverine system by passively floating in the river, many seeds spread in this way as well. When the seeds are stuck at the dams, this has the effects of a fragmentation of the river vegetation. The plants that have seeds that float well will have an advantage over those plants with seeds that do not have as good floating properties. The result is a change in the composition of species along the river.

Although not all hydropower projects require resettlement, involuntary displacement is one of the most sensitive socioeconomic issues surrounding hydropower development (WCD, 2000; Scudder, 2005). It consists of two closely related, yet distinct processes: displacing and resettling people as well as restoring their livelihoods through the rebuilding or rehabilitation of their communities.

When involuntary displacement cannot be avoided, the following measures might contribute to optimize resettlement outcomes:

- Involving affected people in defining resettlement objectives, in identifying reestablishment solutions and in implementing them; rebuilding communities and moving people in groups, while taking special care of indigenous peoples and other vulnerable social groups;
- Publicizing and disseminating project objectives and related information through community outreach programs, to ensure widespread acceptance and success of the resettlement process;
- Improving livelihoods by fostering the adoption of appropriate regulatory frameworks, by building required institutional capacities, by providing necessary income restoration and compensation

programs, and by ensuring the development and implementation of long-term integrated community development programs;
- Allocating resources and sharing benefits, based upon accurate cost assessments and commensurate financing, with resettlement timetables tied to civil works construction and effective executing organizations that respond to local development needs, opportunities and constraints (Kumar et al., 2011).

Finally, some comments about the health impact of the construction and operation of a hydropower plant. In warmer climate zones, the creation of still-standing water bodies such as reservoirs can lead to increases in waterborne diseases like malaria, river blindness, dengue or yellow fever, which need to be taken into account when designing and constructing reservoirs for supply security, which may be one of the most pressing needs in these regions.

In other zones, a temporary increase in mercury may have to be managed in the reservoir, due to the liberation of mercury from the soil through bacteria, which can then enter the food chain in the form of methyl mercury. In some areas, human activities like coal burning (North America) and mining represent a significant contributor.

Moreover, higher incidences of behavioral diseases linked to increased population densities are frequent consequences of large construction hydropower plants. Therefore, public health impacts should be considered and addressed from the outset of the project.

The reservoirs that are likely to become the host of waterborne disease vectors require provisions for covering the cost of health care services to improve health conditions in affected communities. In order to manage health effects related to substantial population growth around hydropower reservoirs, options may include controlling the influx of migrant workers or migrant settlers as well as planning the announcement of the project in order to avoid early population migration to an area not prepared to receive them. Moreover, mechanical and/or chemical treatment of shallow reservoir areas could be considered to reduce the proliferation of insects carrying diseases, while planning and implementing disease prevention programs. Additional options include increasing access to good quality medical services in project affected communities and in areas where population densities are likely to increase as well as establishing detection and epidemiological monitoring programs, establishing public health education programs directed at the populations affected by the project, and implementing a health plan for the workforce and along the transportation corridor to reduce the risk of transmittable diseases (e.g., sexually transmitted diseases) (Kumar et al., 2011).

SOME ADDITIONAL CONSIDERATIONS

The following are a group of additional considerations that EU governments and the private energy sector could considered during the future structure of the energy mix of the different EU Member States.

- Hydropower is a major renewable generation technology in Europe's electricity mix: it is competitive, efficient, climate-friendly, and contributes to system stability;
- Europe's hydropower potential can–and should–be further developed;
- Hydropower improves the Europe's security of supply;
- Hydropower is extremely resource-effective. Hydropower has both a high energy payback ratio and high conversion efficiency, making it very resource-effective. It has the highest efficiency rate among all electricity generating technologies, and offers the highest levels of services;
- Hydropower helps to mitigate climate change. Hydropower is a renewable generating technology with a very low carbon footprint. In addition, thanks to associated water management services like flood and drought control, hydropower can also play a key role in climate change adaptation efforts;
- Hydropower is the only renewable large-scale energy storage technology available today, and represents the most efficient and economical way to store electricity;
- Hydropower's unique services should be optimized on a European scale. Hydropower storage technologies are currently designed for a national or regional scale, with limited consideration for a European system. Since hydropower and pumped storage is scarce and highly valuable resources, their potential has to be used to its optimum on a European or even pan-European scale. Improved grids and interconnections are of prime importance to tap into this significant unused potential of renewable energy generation;

- Obstacles to benefit from more flexible and renewable energy need to be overcome. Despite the considerable advantages of hydropower, several obstacles hamper its further development:
    - Long planning and lead times involving many stakeholders;
    - Complicate administrative procedures and large time periods required for obtaining licenses, concessions and permits for the construction of new hydropower plants;
    - Lack of an appropriate strategic river basin management tool for integrated water use;
    - Trade-offs between conflicting environmental, social and economic objectives;
    - Strong opposition of public opinion to the construction of new large-scale hydropower plants;
    - Lack of harmonized implementation of EU regulation in the energy sector in different national legislations;
    - Grid fees for pumped storage hydropower plants, which ignore its role as an ancillary service;
    - Lack of a reliable long-term investment framework that offers market price incentives to power suppliers, not just to grid operators.
- The unbundling process should continue (RESAP, 2011).

CONCLUSION

Hydropower is one of the renewable energy sources that have good prospects for the future in Europe. While the potential in Scandinavia is well known, it is not the same for Southeastern Europe. This is surprising, since the region still offers considerable untapped, economically viable potential.

All major long-term trends indicate that energy prices, particularly for fossil fuels, are going to continue to increase noticeably in the future. This makes investing in hydropower plants even more of an interesting proposition, bearing in mind that once the initial investment has been made, virtually no further running costs are incurred. This is why the prospect of higher prices for fossil-based electricity sources such as coal, natural gas and oil, could boost the returns on investment in hydropower in the medium-term. The increasing internalization of the external climate costs will lead to increasing burdens on the fossil-fuelled power plants, while zero-carbon hydropower benefits.

Hydropower is also in a better position in comparison with other renewables as a source of electricity. The use of some of renewables such as wind and solar power have to be promoted via incentive programs adopted by interest governments, but in the case of hydropower the use of this type of renewable energy source for electricity generation have been tested for decades and in many quarters such investment is profitable even without government subsidies.

However, the use of hydropower for the generation of electricity could also harbors risks. These include not only technical default risks and landslides, but also variable precipitation or even pronounced drought phases.

Investment in hydropower, in small and medium-sized plants in particular, will probably never be able to generate the extremely high returns known from the days of the dotcom bubble or other short-term excesses. But for investors who are more interested in a stable, albeit relatively somewhat lower return investments in hydropower plants, may certainly provide an optimum mix of returns and stability. Perhaps it is due to the minor scale of small hydropower plants that investors have, over the past few years, failed to discover the many opportunities yet to be seized in the promising, but often still rather remote regions of Europe.

In conclusion, due to political and economic reasons the prediction of the future development of Europe's hydropower park is a difficult task. Public opposition to large hydro schemes will probably result in very little new development of big dams and reservoirs during the coming years. Small scale and low head hydropower capacity will probably increase in the future as research on low head turbines and standardized turbine production, lowers the costs of hydropower plants at sites with low heads. A new computerized control systems and improved turbines may allow more electricity to be generated from existing hydropower plants in the future. As well, many small hydropower plants were abandoned in the 1950s and 1960s when the price of oil and coal was very low, and their environmental impacts unrealized. Increased fuel prices in the future could result in these facilities being refurbished.

It is important to single out that there is a clear agreement or understanding within the EU Member States that a dismantling of the existing hydropower plants is not an advisable option; they should continue to operate, and perhaps their number and capacities will be increased in the future.

References

Auer, Josef (2013). *The new global power plant order. Unconventional and green energies are driving change*; Editor Antje Stobbe; Deutsche Bank am Main; Germany; 2013.

EC (2000). *Energy in Europe–1999 Annual Energy Review*; Special Issue January 2000; Directorate-General for Energy; 2000.

Eurelectric (Union of the Electric Industry) (1997c). *Hydroelectricity, energy vector for progress and development*; Hydro Power and other Renewable Energies Study Committee; EURELECTRIC, Brussels, Belgium; http://www.eurelectric.org; 1997.

Hydro in Europe: Powering Renewables full report (2011). Union of the Electricity Industry–EURELECTRIC; EURELECTRIC Renewables Action Plan (RESAP); 2011.

Hydropower Potential in Switzerland and Europe Usage Today – Future Potential –Perspectives; Swiss Federal Institute of Technology; Zurich, Switzerland; s3.amazonaws.com/zanran.../www.../44115138.pdf.

Jansson, Roland (2006). *The effect of dams on biodiversity*; Department of Ecology and Environmental Science; Umeå University; 2006.

Kumar, A., Schei, T., Ahenkorah, A., Caceres Rodriguez, R., Devernay, J. M., Freitas, M., Hall, D., Killingtveit, Å. & Liu, Z. (2011). *Hydropower.* In IPCC Special Report on Renewable Energy Sources and Climate Change Mitigation [O. Edenhofer, R. Pichs-Madruga, Y. Sokona, K. Seyboth, P. Matschoss, S. Kadner, T. Zwickel, P. Eickemeier, G. Hansen, S. Schlömer, C. von Stechow (eds)], Cambridge University Press, Cambridge, United Kingdom and New York, NY, USA; 2011.

Leonard, J. Crouzet (2000). *Lakes and reservoirs in the EEA area*; European Environmental Agency (2000); Topic report No 1/1999; 2000.

Lehner, B., Czisch, G., and Vassolo, S. (1998). *Hydropower potential today and in the future*; Center for Environmental Systems Research; University of Kassel, Institut für Solare Energieversorgungstechnik (ISET); Kassel, Germany; 1998.

Lehner, B., Czisch, G., and Vassolo, S. (2005*). The impact of global change on the hydropower potential of Europe: A model-based analysis*; Energy Policy, 33 (7), 2005.

Renewables 2013 Global Status Report; REN 21 Steering Committee; ISBN 978-3-9815934-0-2; Paris, France; 2013.

Ruska, Maija & Kiviluoma, Juha (2011). *Renewable electricity in Europe: Current state, drivers, and scenarios for 2020*; VTT Technical Research Centre of Finland; Finland, 2011.

Scudder, T. (2005). *The Future of Large Dams: Dealing with Social, Environmental, Institutional and Political Costs*; Earthscan, London, UK; 2005.

Sørensen, Bent (2000). *Renewable Energy, Its physics, engineering, use, environmental impacts, economy and planning aspects*; Second edition, Academic Press; 2000.

UCTE (Union for the Coordination of Transmission of Electricity) (2000); UCPTE Annuaire Statistique /Statistisches Jahrbuch / Statistical Yearbook 1999; Luxembourg; http://www.ucte.org.; 2000.

UNWWAP (2006). *Water–A Shared Responsibility*; United Nations World Water Development Report 2, World Water Assessment Programme, United Nations Educational, Scientific and Cultural Organization (UNESCO); Paris, France; www.unesco.org/water/wwap; 2006.

Voigtländer, P. & Gattinger, M. (1999). *Potentiale: Wasserkraft*; In: Brauch, H., Czisch, G., Knies, G. (Eds.); Regenerativer Strom für Europa durch Fernübertragung elektrischer Energie; Moosbach, AFES-Press; September 1999 (1), ISBN 3-926979-71-2; 1999.

WCD (2000). *Dams and Development–A New Framework for Decision-Making*; World Commission on Dams, Earthscan; London, UK; 2000.

World Atlas & Industry Guide (2010). International Journal of Hydropower and Dams, UK; 2010.

In: Hydroelectricity and Power Electronics
Editor: Lucille Burke
ISBN: 978-1-63321-728-7

Chapter 3

RISK MANAGEMENT OF HYDROPOWER PROJECTS IN SOUTHWEST CHINA

Wenzhe Tang[1,2*], Zhuoyu Li[1,2], Maoshan Qiang[1,2], Shuli Wang[1,2], Zhen Lei[1,2] and Youmei Lu[1,2]

[1]Institute of Project Management and Construction Technology, Dept. of Hydraulic Engineering, Tsinghua University, Beijing, China
[2]State Key Laboratory of Hydroscience and Engineering, Tsinghua University, Beijing, China

ABSTRACT

There is a rapidly increasing demand for hydropower in China. However, little research has been conducted to systematically investigate the overall aspects of hydropower development risks. With support of the data collected from a fieldwork survey, this study reports the multiple facets of hydropower development risks in Southwest China as perceived by main project participants. All groups have a common view on the criticality of safety, and the groups also have their own priorities, i.e., resettlement of migrants, incompetence of subcontractors, project delay, inadequate or incorrect design, premature failure of facilities, and ecological and environmental impacts are the key risks to clients, whereas quality of work, financial related risks, reputation, and claims and disputes are the main concerns to contractors. A case study of Three Gorges Project

* Corresponding author: TEL +86 10 62794324, FAX +86 10 62782159, Email: twz@mail.tsinghua.edu.cn (W. Tang)

further demonstrates that, hydropower development risks can be effectively managed by encouraging joint efforts of all participants to achieve the project goals, and hydropower in China, which is competitive in energy systems, can play a key role on producing renewable energy, reducing emissions of CO_2, and providing important social/economical benefits. Future hydropower development should emphasize the interactions between project delivery, environmental and social processes to reach appropriate trade-offs among involving stakeholders, by adequately considering the inter-relations between project participants' risks as well as hydropower project's externalities on a broad view.

Keywords: Energy management; Hydropower development; Risk; Partnering; China

1. INTRODUCTION

Demand for electricity in China is increasing at an annual rate of about 10%, and the current installed Chinese national power-generating capacity (792.5 GW) is second only to that of the United States (1032 GW) [1]. During past decades, the share of water-generated electricity as renewable energy has increased from 2% to 21% of the total amount in China, and hydropower development has become a more and more important way to meet the increasing need of energy [2]. Renewable energy is a suitable way to decrease the CO_2 emissions, and fit for the long-term implementation of sustainable energy systems [3-5]. The amount of hydropower resource in China ranks first in the world, but the installed hydropower capacity only takes up 26% of technically exploitable hydropower resource; and the Chinese electrical power development policy is to speed up hydropower development for saving fossil energy resources and decreasing air pollution [6]. As hydropower resource is mainly in Southwest China [7], currently many major hydropower projects are being delivered in this region. Pursuing the objective of an environmental sustainable, competitive, and secure supply of energy should analyze the quantitative extent of risks and their possible causes[8-9], which help to gain insight into the tradeoffs between environmental and economic objectives for appropriate energy resources allocation [10]. Development of hydropower projects deals with not only the risks on building dams, but also the social and environmental impacts from diverting watercourses, which are closely interrelated [8,11-15]. These risks involve a range of participants who have a specific and conspicuous interest in the project outcomes, and there can be a wide divergence of perspectives on allocating and accepting particular risks [16-17]. The lack of synergy between stakeholders is identified as the key obstacle to the implementation and efficient practice of energy management [18]. Thus, the aim of this study is to understand the multiple facets of hydropower project risks, which assists relevant parties to collaboratively manage hydropower development risks for the sake of efficacy. This paper reports the findings of an empirical hydropower industry survey on the importance of project risks as perceived by main project participants, with a case study of Three Gorges Project (TGP) being conducted to further illustrate the findings.

2. Need to Understand Overall Aspects of Risk Management for Hydropower Projects

Many researchers raised the strategy of partnering, which adopts a cooperative philosophy by maximizing the effectiveness of all participants' resources, to mitigate the risks due to misalignments between project participants [19-25]. Rahman and Kumaraswamy concluded that the construction industry is moving towards joint risk management by the application of partnering principles [16]. Tang et al. further revealed the important role of partnering in enhancing risk management, and indicated that partnering helps participants to share added information by the improvement of open communication, which facilitates decision-making to reduce lost opportunities dealing with project risks [22,23]. This can be attributed to: firstly, individuals from various organizations have different experiences and recognitions, who tend to concern more about the risks that are critical to their own organizations, thus they may not always see the overall picture of the project with the risks behind it [17]; and secondly, partnering arrangements are underpinned by incentive schemes, which use clearly defined risks/rewards allocation to manage the processes, providing powerful motivations and necessary resources for participants to achieve project goals [20,21,25,26]. Despite the strategies addressed by the above studies, there has been little systematic analysis of hydropower development risks and the treatment of risks underlying use of incentives. Therefore, it is imperative for all project participants to get to grips with hydropower development risks, and there is a clear need to systematically investigate the overall aspects of risk management on the views of various project participants [6,12,14,16] .

Hydropower project risk management involves a wide variety of factors, including estimations, contracts, clients, contractors, designers, unknown geological conditions, economic and political circumstances, technology issues, natural events, ecological and environmental impacts, affected local communities, management and control techniques, individual activities, organizational operations, and inter-organizational risk allocations [9,11,13-15,17,27-29]. Accordingly, these themes have been systematically investigated in this study, and questionnaire was used as the principal method to collect data, in which the questions were designed based on the above reviewed studies as well as the writers' experience and recognition of hydropower project risks. Besides, the Three Gorges Project has been adopted as a case to interpret the key points from the survey.

3. Empirical Research Methodology

3.1. Data Collection Using a Triangulated Approach

A triangulated approach enables both qualitative and quantitative data collection to be used to test or understand a research proposition [30]. Accordingly, the methods of questionnaire, interview, direct observation, document review and case study were applied for collecting both broad and in-depth data for this study.

Use of questionnaire was chosen as the principal survey method. Questions were formatted using a five-point Likert scale, and this format permits different statistical techniques to be used to analyze the data. Postal surveys are beneficial in avoiding excessive leg-work, but amplify the problem of obtaining an adequate level of response [31]. Akintoye and Macleod argued that postal surveys may be biased if the return rate is lower than 30-40% [32]. To avoid these limitations, a questionnaire survey was conducted through fieldwork. Six field trips to hydropower project sites in Southwest China were undertaken. Southwest China was chosen as the study area because 70.8% of hydropower resources are located in this region [7], with many major hydropower projects already in the processes of delivery and operation, providing a rich source of data for this study. The total economically exploitable amount of hydropower resource in China is 402000 MW. This study surveyed 12 hydropower projects, with the installed capacity of these projects (81400 MW) accounting for 20% of the above total amount.

To understand the overall issues in hydropower project risk management, the key participants of the projects were used as respondents, including clients, contractors (Contr.), designers (Desi.), and superintendents (Super.). Superintendents have dual roles: to act as agents for clients in conveying the clients' instructions to the contractors and to act as certifiers for the purpose of issuing certificates and making decisions as to reasonable measures of the value of work, quantities or time. The respondents were chosen because they had experience in the delivery of significant projects, such as the Three Gorges Project. A total of 268 questionnaires were collected from hydropower project sites in the Sichuan, Yunan, Chongqing and Hubei provinces in Southwest China. Among the respondents, 30% held senior positions in their organizations, such as director, general manager, project manager, department head and chief engineer; the remaining respondents had also been involved in the hydropower industry for many years. The chosen respondents were initially identified via personal relationships and reputation. Direct contact confirmed their willingness to assist in this study. The fieldwork approaches used in this survey enabled all sent questionnaires to be collected, thus the response rate reaches 100%.

The fieldwork in this study was conducted such that, after the questionnaires had been completed, interviewing with the respondents holding senior positions were conducted. Direct observation during site visits also enhanced the researchers' understanding of the hydropower project environment and construction processes, which are closely related to various risks. Some project documents were reviewed to help elucidate how risks arise. In addition, a case study of Three Gorges Project was conducted for in-depth analysis. The data collected through the above methods were used to test and interpret the results from the questionnaire survey. Given the geographic distribution of hydropower resources, the hydropower project development progress, the variety of respondents, and the triangulated data collection approach, the bias of selecting samples can be reasonably reduced, and the data collected can, to a large extent, be taken to be representative of the hydropower industry in Southwest China.

3.2. Data Analysis Techniques

The data collected from questionnaires were analyzed using the Statistical Package for Social Science [32]. The selected techniques for this study include estimation of the sample population mean, rank cases, one-way analysis of variance (ANOVA), and Spearman rank correlation. Of these statistical techniques, the ANOVA and Spearman rank correlation were adopted for inferential analysis, with the results being tested for significance. The level for significance in this study follows the usual level for statistical significance of 0.05, with a level of 0.01 being highly significant.

4. SURVEY RESULTS AND ANALYSIS

4.1. The Importance of Hydropower Project Risks

Respondents were asked to rate the importance of 55 possible risks identified in the literature, listed in the first section, on a scale of 1-5, where 1 represented negligible risk and 5 extreme risk. The results are presented in Table 1.

Overall, the five most important risks are “safety”, “incompetence of subcontractors”, “quality of work”, “premature failure of facilities”, and “delay”. One-Way analysis of variance (ANOVA) was performed to test the perceptions of clients, contractors, superintendents and designers. The One-way ANOVA is used to test whether several means are equal [33], and the results show that the groups have different perceptions on the importance of some risks.

To test whether there was consensus among the various groups on the rankings of the importance of project risks, the Spearman rank correlation coefficient r_s was computed, as shown in Table 2.

In this table the top entry in each cell is the Spearman correlation coefficient, and the lower entry is the significance level (p-value). The Spearman rank correlation coefficients r_s between groups show that the risk ranking of clients has no significant agreement with the risk rankings of contractors, superintendents and designers. Contractors’ ranking also has no significant agreement with designers’ ranking. Only superintendents have significant agreement with contractors and designers at the $p < 0.01$ level, which can be attributed to superintendents’ role of in charging whether contractors’ performance meets the requirements of designers at the project implementation stage. This situation shows that the priorities to some risks for the key project participants are different from each other. As project risks are mainly allocated to clients and contractors, it is necessary to further discuss the priorities on the same risks for these two groups as below.

Table 1. Perceptions of respondents on importance of risks

Risks	Overall		Client		Contr.		Super.		Desi.	
	M.	R.	M.	R.	M.	R.	M.	R.	M.	R.
Safety	3.409^{b}	1	3.175	5	3.395	3	3.80	1	4.07	1
Incompetence of subcontractor	3.301	2	3.277	2	3.299	17	3.45	8	3.06	42
Quality of work	3.285	3	2.938	8	3.558	1	3.63	2	3.38	22
Premature failure of facilities	3.255	4	3.116	6	3.267	19	3.63	3	3.19	33
Delay	3.227	5	3.211	3	3.303	14	3.19	24	3.06	45
Inaccurate marketing prediction for future growth	3.173^{b}	6	2.866	17	3.360	9	3.38	12	4.00	3
Inadequate core capabilities of organization	3.162^{b}	7	2.866	18	3.187	25	3.54	5	4.00	4
Cost control in project implementation	3.152^{b}	8	2.814	22	3.487	2	3.23	21	3.75	12
Lack of motivation for staff	3.145^{b}	9	2.786	26	3.250	21	3.60	4	3.94	5
Lack of innovation	3.124^{b}	10	2.848	19	3.227	23	3.45	9	3.86	6
Resettlement of migrants	3.102^{b}	11	3.553	1	2.613	55	2.81	51	3.13	40
Inadequate or incorrect design	3.087	12	3.204	4	2.907	45	2.98	43	3.50	18
Shortage of skills/techniques	3.087^{b}	13	2.748	29	3.338	11	3.53	6	3.06	41
Inappropriate organizational response to changing environment	3.067^{b}	14	2.893	14	3.053	36	3.19	26	4.00	2
Lack of organizational learning capability	3.052^{b}	15	2.739	31	3.162	26	3.40	10	3.81	7
Inappropriate organizational structure	3.047^{b}	16	2.714	35	3.234	22	3.35	14	3.75	11
Ineffective integrating and managing enterprise resources	3.027^{b}	17	2.652	39	3.276	18	3.40	11	3.56	14
Incapability of making profits	3.024^{b}	18	2.634	42	3.395	4	3.19	25	3.56	13
Inefficiency of business processes	3.016	19	2.893	15	2.987	42	3.21	23	3.50	17

Table 1. (Continued)

Risks	Overall		Client		Contr.		Super.		Desi.	
	M.	R.	M.	R.	M.	R.	M.	R.	M.	R.
Delay of drawing supply	3.008	20	2.929	9	3.316	12	2.92	48	2.50	55
Insufficient technology	3.004[b]	21	2.655	38	3.373	8	3.23	22	3.06	43
Reputation	3.004[b]	22	2.536	50	3.387	6	3.48	7	3.19	30
Material or equipment quality	3.004	23	2.867	16	3.147	28	3.04	39	3.31	25
Claims and disputes	2.996[b]	24	2.661	37	3.377	7	3.08	32	3.31	26
Unclear strategy for achieving organizational objectives	2.992[a]	25	2.770	28	2.987	41	3.29	16	3.75	9
Poor coordination	2.992[a]	26	2.804	24	3.078	33	3.16	27	3.38	21
Failure to identify defects	2.992	27	2.920	10	3.026	39	3.08	30	3.19	34
Poor financial situation	2.984[b]	28	2.652	40	3.395	5	3.04	35	3.19	31
Feasibility of construction method	2.984	29	2.821	21	3.132	29	3.08	31	3.13	38
Shortage of labour, materials and equipment	2.969[a]	30	2.735	33	3.303	15	2.96	45	3.00	47
Mismatch between organizational strategy and culture	2.968[b]	31	2.709	36	2.974	44	3.38	13	3.56	15
Inadequate project management methods	2.965[a]	32	2.739	32	3.039	38	3.25	19	3.31	23
Inadequate market share	2.952[b]	33	2.634	43	3.257	20	3.27	17	2.81	54
Site access	2.949	34	2.726	34	3.311	13	2.96	44	2.94	48
Poor relations with external organizations	2.945	35	2.786	27	2.987	43	3.23	20	3.00	46
Difficulty in entering new business	2.937[b]	36	2.545	49	3.132	30	3.27	18	3.75	10
Third party delay	2.925	37	2.814	23	3.091	31	2.89	49	2.94	50
Unforeseen site condition	2.918	38	2.912	11	2.895	46	3.00	41	2.88	53
Ineffective communication	2.910[b]	39	2.595	46	3.091	32	3.08	29	3.81	8

Risks	Overall		Client		Contr.		Super.		Desi.	
	M.	R.	M.	R.	M.	R.	M.	R.	M.	R.
Force majeure	2.909	40	2.894	13	2.767	51	3.04	38	3.31	24
Incompetence of products and services	2.901[b]	41	2.369	54	3.360	10	3.34	15	3.25	27
Quantity variations	2.897	42	2.902	12	2.855	48	2.94	46	2.93	52
Ecological and environmental impacts	2.895	43	3.062	7	2.711	53	2.79	52	2.94	51
Inappropriate risk allocation	2.878[b]	44	2.646	41	3.197	24	2.70	53	3.44	19
Exchange rate fluctuation and inflation	2.870	45	2.788	25	3.053	35	2.67	54	3.20	29
Organizational growth	2.851[b]	46	2.571	47	3.000	40	3.04	34	3.50	16
Poor relationship between parties	2.847[b]	47	2.566	48	3.066	34	3.04	40	3.13	39
Organizational interface	2.842	48	2.741	30	2.865	47	3.04	37	2.94	49
Change in codes & regulations	2.837	49	2.830	20	2.737	52	2.92	47	3.14	35
Poor definition of scope	2.835[a]	50	2.407	53	3.303	16	3.00	42	3.06	44
Logistics	2.807[b]	51	2.611	44	3.158	27	2.55	55	3.13	37
Inadequate planning	2.792[a]	52	2.602	45	2.816	50	3.06	33	3.19	32
Other risks due to uncertainties on government policies and economy	2.723[a]	53	2.411	52	2.830	49	2.82	50	3.42	20
Adequacy of insurance	2.722[b]	54	2.248	55	3.039	37	3.13	28	3.25	28
Conflicts in documents	2.689[b]	55	2.509	51	2.632	54	3.04	36	3.13	36

Note： M.=mean; R.=rank; [a] ANOVA is significant at the 0.05 level; [b] ANOVA is significant at the 0.01 level.

Table 2. Correlation among participants on importance of risks

	Client	Contr.	Super.	Desi.
Client	1.000			
	.			
Contractor	-0.077	1.000		
	0.575	.		
Superintendent	0.060	0.463 [b]	1.000	
	0.662	0.000	.	
Designer	-0.073	0.140	0.496 [b]	1.000
	0.597	0.309	0.000	.

Note: [b]=correlation is significant at the .01 level (2-tailed).

4.2. Clients and Contractors' Common Concern: Safety

Both clients and contractors are very concerned about "safety" and ranked the risk fifth and third respectively, which can explain that "safety" is the most important risk in overall. To understand the safety incidents' impacts, an insurance company's safety incidents records on hydropower project delivery during past three years were studied. The frequency of safety incidents and the number of deaths caused by these incidents are shown in Table 3.

Table 3. Impacts of safety incidents in hydropower project delivery

Types of incidents	Frequency of incidents	Number of deaths
Collapses of underground works	8	36
Landslides	6	31
Mudflows	3	75
Lifting accidents	3	13
Transportation accidents	3	13
Struck by falling objects	2	8
Explosions	2	8
Fall to lower level	1	3
Collapses of working support systems	1	3
Total	29	190

Among the 29 safety incidents during past three years, 8 incidents were "Collapses of underground works". These incidents were most frequently happened, including failures such as collapses of bases and tunnels, which caused deaths of 36. The causes of these failures involve insufficient planning, unforeseen geology conditions and inappropriate construction methods. "Landslides" happened 6 times, which is the second most frequently happened risk, resulting in deaths of 31. The consequences caused by landslides were attributed to failing to predict the influence of construction processes on the stability of lands, and insufficient reinforcement to the steep slopes. "Mudflows" happened 3 times, but they resulted in the most deaths of 75. Hydropower projects are normally set in valley with steep slopes. When extreme weather events occurred, floods together with soil erosions and landslides can result in mudflows, which cause significant impacts if there is inadequate early warning. "Lifting accidents", "transportation accidents", "struck by falling objects", "explosions", "fall to lower level" and "collapses of working support systems" also resulted in deaths, which were relevant to personal skills, safety training, site conditions, safety supervision in construction, overtime work due to tight construction schedules, protective equipment, and safety culture.

4.3. Clients' Main Risks

"*Resettlement of migrants*" is the most important risk to clients due to the inundation of lands caused by hydropower development. Interviews with senior managers of clients confirmed that "resettlement of migrants" has become a main restriction for hydropower project development. Migration problems can influence the attainment of government approvals, disrupt the construction processes, impede the normal operations of completed hydropower projects, and ultimately result in heavy financial losses to clients. There is a variety of causes for migration-related problems, including: 1) The compensation is too low to cover the losses of migrants due to outdated laws and codes; 2) Government agencies are in charge of relocating migrants, but they need the financial support of the clients, and the financial and management misalignments between government agencies, clients and migrants have increased the difficulty in solving migration problems; 3) Migration planning is insufficient due to the short planning period offered by clients and the technical difficulties in identifying appropriate relocation lands, which frequently lead to variations in developing migration communities at the implementation stage; 4) There are inadequate measures for supporting migrants' sustainable development. Migrants' skills tend to be suited to their original living environment and may be ineffective in new communities. Moreover, migrants have lost the original social networks that form an important part of their social capital, which can further aggravate their economic marginalization and poverty.

"*Incompetence of subcontractors*" is the second most important risk to clients. Due to fierce market competition, many contractors chose a low-price bidding strategy and subcontract most of the work to subcontractors with lower prices. This causes subcontractors to hire only unskilled and semi-skilled workers. Interviews confirmed that approximately 70% of the main contractors' workers are from subcontractors and that these workers are without adequate technical skills. Ultimately, there are uncertainties about subcontractors' technical qualifications, quality of work, timeliness and safety. Although workers are required to undergo technical training before working, the performance of the workers is still worrisome. This occurs because workers are mainly from rural communities with low educational levels, which significantly affects their skill improvement. Normally, the jobs of the workers are *ad hoc*, and the remuneration of workers is low. These conditions result in instability in the work forces, which also impedes the continuous skill improvement among workers. The dilemma of workers from rural areas regarding their social disadvantages in education and employment is a nation-wide issue. These problems require the joint efforts of clients and contractors at the project management level and systematic institutional improvements in the long run at the national level.

"*Project delay*" is ranked as the third most important risk by clients. Delays can result in extra construction costs for clients. More importantly, project delays can cause a large income reduction due to the delay of electricity generation. Interviews showed that the main causes of delay are largely related to other key risks of clients, ranging from the first to the eleventh items listed in Table 1. "Resettlement of migrants" (first) and "ecological and environmental impacts" (seventh) reflect hydropower project externalities, which could result in significant delays. For example, in one case, the normal electricity generation of a major project was put off for three years due to disputes between clients and local migrants. "Inadequate or incorrect design" (fourth) and "delay of drawings" (ninth) could cause serious project delays due to the designers' key role in technical issues. "Unforeseen site conditions" (eleventh) leading to variations are a frequently appearing cause for delays at the implementation stage. Project performance-related risks, such as "incompetence of subcontractors" (second), "safety" (fifth), "premature facility failure" (sixth), "quality of work" (eighth) and "failure to identify defects" (tenth) are also important causes of delay during project construction. Thus, preventive actions dealing with project delay not only should focus on proper time estimation and programs but also should aim to reduce the other key risks to clients.

"*Inadequate or incorrect design*" was ranked fourth by the clients. Interviews confirmed that many problems, such as rework, delay, additional costs and claims, originate from design-related factors. The causes of inadequate or incorrect design involve market, technical and client requirements. Firstly, due to the increasing need for energy, many more hydropower projects are being developed than before, and designers lack the resources needed to undertake an in-depth study of the projects, resulting in inadequate or incorrect design. Secondly, many hydropower projects have high dams and complex geologic conditions, which could pose significant technical challenges to the designers. Thirdly, "inadequate or incorrect design" is also related to client strategy. Once a project has been approved by the government, the client tends to push designers to complete the detailed design as soon as possible, which could result in inadequate design and frequent variations in the work during project implementation. Thus, to deal with design risks, both clients and designers should promote open communication for clarifying clients' intentions and discussing problems that designers may encounter. Such discussions are essential for making the changes needed for hydropower development.

"*Premature failure of facilities*" is ranked sixth by the clients. Premature failure of facilities during hydropower project construction processes have a variety of causes, which can be incorrect design, inadequate planning, poor implementation, and defects of material and equipment. Besides, incidents attributed to extreme weather and unforeseen geological conditions such as floods, landslides and mudflows can seriously destruct the facilities of ongoing projects. The above causes to "premature failure of facilities" are also closely related with "safety".

"*Ecological and environmental impacts*" is ranked seventh by the clients. The construction of dams typically results in habitat loss and changes to river flows; these changes act to reduce biodiversity, such as the diversity of fish species [Palmer 2010, 14]. However, clients and designers traditionally tend to focus on engineering issues, resulting in a lack of in-depth study of dams' ecological and environmental impacts at an early stage. Clients are increasingly recognizing the importance of adequately studying, in advance, the relationships between dam construction, ecological/environmental processes, and social factors in hydropower project development. Understanding these linkages can help to balance the needs of humans and nature to avoid high environmental, economic and social costs and can assist in reaching appropriate trade-offs between competing stakeholders due to the changes in ecological and environmental systems caused by hydropower projects.

4.4. Contractors' Main Risks

"*Quality of work*" is the most important risk to contractors. It is well recognized by top-level contractor managers that the survival of a contractor largely relies on the quality of services and projects supplied to clients. Contractors have established quality systems and processes, but the levels of effective implementation need to improve in hydropower project delivery. A key barrier to quality management is the fact that the workforce generally has a low educational level and inadequate technical skills, which have impeded the effectiveness of training and the implementation of quality systems, processes and continual performance improvement. This workforce problem is consistent with the clients' second most important risk of "incompetence of subcontractors". Another key barrier is the fact that main contractors normally keep only a very limited experienced management staff due to financial pressure, which also has weakened the contractors' quality management capability. In addition, a lack of in-depth communication between project participants can lead to quality-related problems and disputes. This is not surprising because the performance of any individual organization is largely dependent on its interactions with other organizations in terms of the complexity of hydropower projects.

"*Cost control in project implementation*", "*capability of making a profit*", and "*poor financial situation*" are ranked as the second, fourth and fifth most important risks by contractors respectively, whereas they are ranked as the 22nd, 42nd, and 40th risks respectively by clients, showing that the clients' financial situation is much better than that of contractors. Interviews confirmed that the financial difficulties of contractors can largely be attributed to the traditional lowest responsive bidder selection process, which places extreme pressure on the contractors to provide a marginally adequate bid to cover the work. For example, an interviewed contractor is facing significant financial problems due to unforeseen geologic conditions that should have been adequately considered in advance to implement an underground work. The interviewed contractors indicated that they had been forced to concentrate more on survival than on investing in improvements, such as hiring skilled workers, renewing equipment and developing competent technologies. This situation can further affect contractors' ability to make a profit, which leads contractors to fall into a vicious cycle of development.

"*Reputation*" is ranked sixth by the contractors, indicating that a good reputation is also an important objective for contractors after their completion of a project. This result arises because a reputation is essentially an overall reflection of the contractors' capabilities with respect to technology, equipment, management, finances, human resources, learning and innovation, and client relationships. A good reputation is a pre-condition to win the trust of clients, which can facilitate contractors in obtaining scarce resources and in expanding their market share. Interviews confirmed that the contractors who won the bids of hydropower projects in Jinsha River, which are being developed by the Three Gorges Corporation, were largely chosen based on the reputation established in their successful construction of the TGP. Considering the importance of reputation, contractors should appropriately balance their objectives between making profits and enhancing other capabilities.

"*Claims and disputes*" is ranked seventh by the contractors. Due to the complexity of hydropower projects, many problems and issues have arisen at interfaces, which could lead to claims and disputes between the involved parties. "Inadequate or incorrect design", which frequently leads to project variations, changing programs and reworks, is also an important source of claims and disputes. Comparatively, clients ranked "claims and disputes" as the 37th risk and consider them to be normal activities that should be within the coverage of the contingencies. This indicates that clients are much more capable of bearing risks than contractors. Interviews confirmed that most clients are able to use a contingency to cover unforeseen risks, but it is difficult for contractors to cover serious risk incidents because of their narrow margins. Because the impact of "claims and disputes" on finances for clients is not as great as it is for contractors, there are few significant conflicts between clients and contractors on sharing the risks on most issues, which provides a sound base for clients and contractors to resolve disputes jointly.

Analyzing the causes of the above clients and contractors' risks demonstrates that one group's risks are largely interrelated with other groups' risks, e.g., "project delay" can be due to a variety of risks, confirming the need for participants to jointly manage the hydropower development risks by partnering approach. A case study of TGP was conducted to further illustrate this as below.

5. Case Study of Three Gorges Project

5.1. Risk Management Strategies in Three Gorges Project

The Three Gorges Project is the largest hydropower project in China. The project plays a key role in harnessing the Yangtze River and developing the adjacent regions, which is located in Sandouping town, Yichang, Hubei province. The project comprises dams, navigation facilities, and power plants with installed capacity of generators being 18200 MW. The key functions of the project are electricity generation, navigation and flood control, and the total investment budget is RMB 203.9 billion (US $32.8 billion) with the planned project delivery period from 1993 to 2009. The main participants of the project are the client - Three Gorges Corporation, the designer - Changjiang Water Resources Commission, superintendents and contractors. Among them, the client takes charge of the TGP development by integrating and managing the implementation activities that involve the project participants. Generally, the risk management strategies adopted in the TGP development processes are consistent with the survey results on risk importance. The features of the TGP risk management practice are in line with the win-win philosophy of partnering that stress joint efforts of project participants by equitably allocating the project rewards/risks, which are illustrated below.

The survey results show that the most important risk is “safety”, which is the common concern of project participants (see Table 1). As safety incidents were largely attributed to complex conditions of geology (see Table 3), a total of RMB 11.3 billion was spent on the prevention and treatment of geological hazards in the reservoir area. In the 646 affected communities, 69900 residents were relocated; 441 locations prone to landslide and 2874 high-cut slopes were technically treated; 175km of reservoir bank protection was provided; 3113 public monitoring points involving 600 thousands of residents were set up for early warning.

The cost spent on the TGP migration can illustrate hydropower projects’ social impacts and the importance of “resettlement of migrants” to clients (see Table 1). At the water impoundment level of 175m, the TGP reservoir has a length of 650km, a surface area of 1084 km^2, and an average surface width of merely 1.1 km. The reservoir has inundated 632 km^2 of land, affecting 20 counties and cities. It was forecasted that 1.13 million migrants should be resettled, with relevant costs accounting for 44.4% of the total budget. During the course of 17 years of project development, 1.25 million affected residents have been relocated, and the share of resettlement costs has increased to about 50% of the actual project costs. The approaches of resettlement include hiring relocation lands, building new communities, developing local infrastructure, mitigating geological hazards, compensations, and improving migrants' skills for sustainable development.

The survey results demonstrate that the priorities to some risks for the key project participants are different from each other (see Table 1). To align the objectives between the client and contractors, a combined incentive scheme based on win-win value of partnering was adopted by the client to ensure that contractors have the motivations and necessary resources to carry out the tasks assigned in the construction processes of TGP. The incentive schemes typically include incentives on: quality (Weight = 45%), schedule (Weight = 24%), occupational health & safety (OH & S) and environment (Weight = 15%), information management (Weight = 10%), and coordination (Weight = 6%), which had various measures to allow an appropriate rewards/risks sharing to be decided. The value of rewards from incentives can reach 3% of the cost of a work, which could effectively assist contractors in improving their financial situation by good performance.

The weights of the incentives indicate the importance of a risk that a particular incentive deals with as perceived by the project participants. Consistent with the survey results showing that "quality of work" has been identified as most important to the contractors (see Table 1), the quality incentive in the TGP is given the heaviest weight. It is recognized that quality of work is relevant to all project participants, and the TGP quality management committee has been formed to achieve the quality objective that aims at zero quality failures, which include the key members from the client, the designer, contractors and superintendents. In line with "project delay" being ranked the third important risk by clients in the survey, the schedule incentive is set in the TGP to ensure the progress of the project construction is under control. OH & S and environment is also identified as a key issue in the TGP construction, which is consistent with the survey results on "safety (ranked first)", "premature failure of facility (ranked fourth)" and "ecological and environmental impacts (ranked seventh by clients)" as shown in Table 1, thus the relevant incentive is set to mitigate this risk. Use of incentives in the TGP encourages all participants to collaboratively manage key project risks by reducing the likelihood of occurrence and the consequences of them. For example, comprehensive measures to reduce the likelihood of quality deviations have been incorporated in the quality incentive scheme, and there are five preventive measures incorporated in the OH & S and environment incentive scheme. The incentive measures to reduce the consequences of risk are also incorporated in the quality, OH & S and environment incentive schemes. When quality, OH & S and environment incidents occur or there are early warnings of problems, these incentive measures require participants to take prompt action to reduce and control the consequences of the risks. In addition, quality, OH & S and environment incentives include measures to jointly investigate and analyse the causes of occurred incidents, and accordingly to upgrade the risk prevention strategies for avoiding the occurrence of similar risks.

Due to the complexity of the TGP with multi-disciplinary components, dealing with above risks during construction processes needs the support of adequate information and coordination for participants to make appropriate decisions; therefore, information management and coordination incentives have been set to enhance joint management of above risks by facilitating open communication among participants. The coordination incentive incorporates clear measures for efficient problem resolution, which stresses all project issues dealt collaboratively. For example, because many problems and issues have arisen from interfaces and one party cannot solve the problems alone in the circumstances, interface management measures are set to facilitate the resolution of conflicts among different jobs, departments and organizations in the interfaces. To improve the efficiency of information circulation, the technical support for improving information management capabilities in the TGP is the development and use of the Three Gorges Project Management System (TGPMS). The TGPMS incorporates computer technologies via communication, database, multimedia, office system, and management system, to deal with issues on quality, scheduling, safety, engineering, procurement, contracts, financing, documents, material and equipment, payment and cost control. TGPMS provides participants with a variety of communication methods to share and manage the large amount of information derived from the construction activities. To improve the effectiveness of TGPMS, the information management incentive clearly specifies the measures regarding the use of TGPMS. Contractors, which use TGPMS efficiently and input all necessary data into the system in a timely manner, are able to obtain rewards. In this situation, project data/information are exchanged correctly and in a timely way among departments, different management levels and organizations, including clients, contractors, designers and superintendents, to enable them to effectively manage the project risks by monitoring, analyzing, and responding to the project activities.

5.2. Outcomes of Three Gorges Project development

The outcomes of the TGP development shows that the TGP risk management strategies are effective in managing the identified key hydropower project risks. The total cost of the project is RMB 173.1 billion (US$ 27.8 billion), which is 15% less than budget of RMB 203.9 billion (US$ 32.8 billion); The project development progress had strictly met the project delivery schedule of 17 years; The safety incidents decreased continuously; The relationships among involved project stakeholders were good; The specified project quality was achieved in a systematic and orderly manner, and the completed works have met the required standards that ensure the TGP's main functions of power generation, navigation and flood control. The amount of power generated by the TGP is shown in Figure 1.

The planned amount of power generated by the TGP is 84.68 billion kwh, equivalent to the amount of power produced by burning about 30 million tons of standard coal. The results in Figure 1 demonstrate that the objective of power generation can be reached since 2010 after the normalization of water impoundment operations. Normally, the cost of renewable energy is relatively high [34], whereas the electricity price generated by the TGP (RMB 0.25 per kwh) is much cheaper than usual electricity price generated by thermal power projects (RMB 0.35 per kwh). This demonstrates that hydropower development in China can play an important role to satisfy the abatement of the CO_2 emission.

It was forecasted that, after the completion of TGP, the one-way shipping capacity can be increased from about 10 million tons to 50 million tons annually. The actual change of shipping capacity is shown in Figure 2.

With the 660 km-segement of the upstream navigation conditions being improved, 10,000-tonnage barge fleets can sail directly from Wuhan to Chongqing, and since 2011, about 100 million tons of goods passed through the TGP per year as shown in Figure 2. The unit towing capacity of ships in the reservoir area has increased from 1.5 t/kwh to 4~7 t/kwh, and the average energy consumption rate of shipping has reduced by 62%, from 7.6kg/1000tkm prior to reservoir formation to present 2.9kg/1000tkm. The cost of waterway transport is RMB 0.026 per tkm, which is lowest in the TGP area, and the proportion of waterway, railway and highway transport costs in the reservoir area is 1 : 5 : 15. During the period from 2003 to 2010, waterway transport has saved 24.55 million tons of fuel due to the improvement of navigational conditions, led to a decrease of 7.365 million tons of CO_2 emissions, 98,000 tons of SO_2 emissions, and 130,000 tons of nitric oxides.

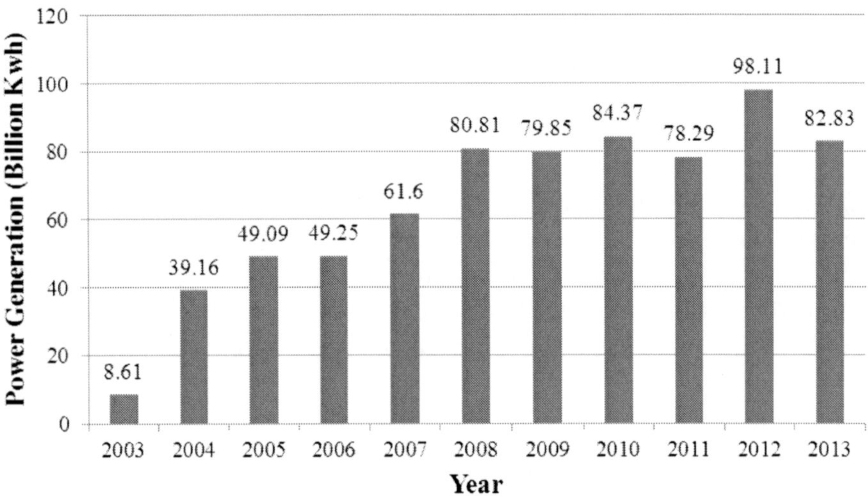

Figure 1. Power generation of the Three Gorges Project.

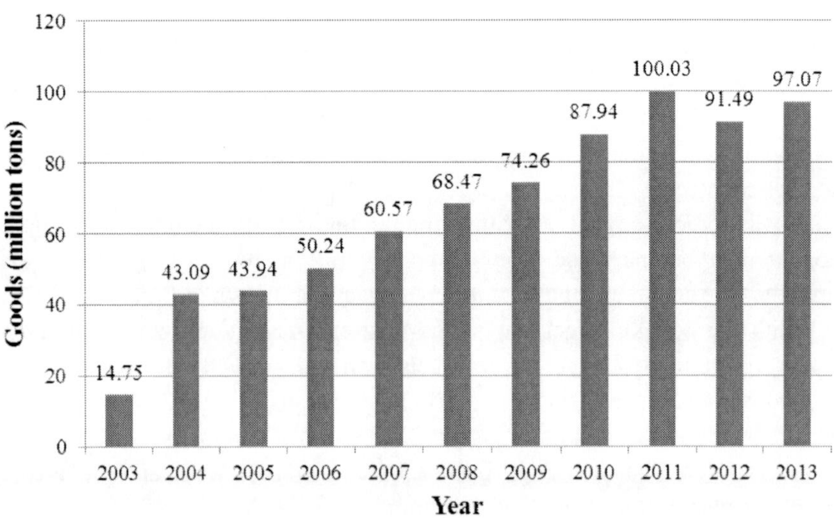

Figure 2. Goods weight of passing through the Three Gorges Project.

The TGP's normal water storage level is at 175m, and the flood control level is at 145m, with a flood-control reservoir capacity of 22.15 billion m^3. As estimated by the Office of the Yangtze River Flood Control Headquarters, the benefits of the TGP had an economic value of RMB 26.63 billion in the flood season of 2010. For example, the highest flood crest reduced was 30000 m^3/s on July 20, 2010, which had largely reduced the pressure of flood control in downstream areas.

6. Practical Strategies on Systematically Managing the Risks of Hydropower Projects in Southwest China

6.1. Promoting Collaborative Risk Management System

The survey results, together with the case study of Three Gorges Project, demonstrate that one group's risks are typically interrelated with other organizations' risks, confirming the need for participants to systematically manage the hydropower project risks by collaborative approach. Collaborative risk management system should allow an organization to measure its risk management system against relevant organizations, which provide comparative information and peer review feedback for the organization to take appropriate actions. Clients, contractors, designers, superintendents and other stakeholders need to establish constructive dialogue involving input to each other's risk management processes, in which participants are able to disseminate project information during reciprocally interdependent processes of hydropower projects planning, design, construction, and operation.

As clients dominate the resources of the whole business chain from hydropower projects during their life cycles, clients play a key role in promoting the collaborative risk management system by adopting incentive schemes such as in the Three Gorges Project. Incentive schemes should follow the win-win philosophy of partnering, which equitably allocate the rewards/risks of hydropower project development to align the objectives among involved stakeholders, ensuring they have the motivations and necessary resources to carry out the tasks for achieving the project objectives of quality, time, cost, safety, migration and environment.

6.2 Expanding Safety Management Scope

As there is a wide range of causes to safety incidents, safety management scope of hydropower projects should be expanded. Due to the feature of hydropower projects' natural environment and complex geology conditions, sufficient project planning is essential to consider the influences of extreme weather and construction processes on the stability of lands, thereby choosing suitable living districts for local residents and workers, taking adequate reinforcement measures to the steep slopes, and setting early warnings for natural hazards. Improving workers' technical skills, safety training, protective equipment at sites, safety supervision in construction, and safety culture building also needs to be emphasized. In addition, because hydropower development normally has tight construction schedules, overtime work should be avoided for the sake of safety.

6.3. Enhancing Capabilities of Workers

Because workforce lacking technical skills is an important cause to the key project risks on "safety", "incompetent subcontractors", and "quality of work", both project management and institutional approaches should be developed to improve workers' capabilities. At the project level, adequate education and training needs to be conducted to enhance the workers' learning technical skills, safety knowledge and following the requirements of project programs, which remove the barrier that most workers from rural areas are with no educational level. At the national level, systematic institutional improvements are required to eliminate the root causes of the social disadvantages for workers from rural areas on education, employment, social security and rewards sharing from project development.

6.4. Sustainable Hydropower Development by Adequately Considering the Interactions between Project Delivery, Environmental and Social Processes

Resettlement of migrants has become a main restriction for hydropower project development, and approaches on legal, governance, migration planning and livelihood of migrants should be taken, which include updating compensation standards by renewing relevant laws and codes, aligning the objectives of clients, migrants and governments by improving the governance systems, sufficient migration planning, and measures to support migrants' sustainable development.

Ecological and environmental impact is another key restriction for developing hydropower projects. Both clients and designers should input more on in-depth study of dams' ecological and environmental impacts at early stage, rather than traditionally only focusing on engineering issues. Understanding the interactions between project delivery, environmental and social processes can help to balance the needs of humans and nature to avoid high environmental, ecological and social costs caused by building dams.

Hydropower resource is mainly in Southwest China with the environmental and social costs borne by this region. However, the benefits of using hydropower mainly go to coastal areas, which purchase the electricity generated by hydropower projects such as the TGP with low price. Initiatives at national level should be adopted to equitably allocate the benefits/costs of hydropower development, e.g., coastal areas pay more to compensate the affected migrants and environment in Southwest China.

CONCLUSION

Based on the perceptions of the main project participants, the survey conducted in this study has systematically revealed the status of the risk management of hydropower projects in China. All groups have a common view on the criticality of safety, which is largely attributed to unforeseen conditions of geology, with collapses of underground works, landslides and mudflows having resulted in the most severe impacts. In addition to safety, the key risks to clients are resettlement of migrants, incompetence of subcontractors, project delay, inadequate or incorrect design, premature failure of facilities, and ecological and environmental impacts, whereas the main concerns to contractors are quality of work, financial related risks, reputation, and claims and disputes. Analyzing the causes of the risks demonstrates that one group's risks are typically related to other organizations' risks, e.g., project delay can be due to a variety of risks, confirming the need for participants to collaboratively manage the risks of hydropower development.

The case study of TGP demonstrates that, although there is a wide range of risks, hydropower development can be effectively managed by partnering approaches such as use of incentives to provide powerful motivations and necessary resources for participants to achieve project goals. The outcomes of the TGP confirm that hydropower in China, which is competitive in energy systems, can play a key role on producing renewable energy, reducing emissions of CO_2, and providing important social/economical benefits such as flood control and navigation.

The practical strategies from above findings suggest that future hydropower development in China should: 1) promote collaborative risk management system to provide comparative information and peer review feedback for participants taking appropriate actions during reciprocally interdependent processes of hydropower projects planning, design, construction and operation; 2) expand safety management scope by sufficiently considering hydropower projects' natural environment, geology conditions, project planning, construction processes, choice of living districts for local residents and workers, treatment to the steep slopes, early warning for natural hazards, protective equipment at sites, workers' education and training, safety supervision in construction, overtime work due to tight construction schedules, and building of safety culture; 3) enhance capabilities of workers to reduce the key project risks on safety, incompetent subcontractors and quality of work by adopting both project management and institutional approaches; 4) reach appropriate trade-offs among involving stakeholders according to the interactions between hydropower project delivery, environmental and social processes.

ACKNOWLEDGMENTS

Many thanks are given to the National Natural Science Foundation of China (Grant Nos. 51379104, 51079070, 51179086, 50539130, 70671058), State Key Laboratory of Hydroscience and Engineering (Grant Nos. 2013-KY-5, 2009-ZY-7), and Education Ministry of China (Grant No. 20091011294). Special thanks are also given to the respondents for their generous contributions during the survey.

REFERENCES

[1] McElroy, MB; Lu, S; Nielsen, CP; Wang, Y. Potential for wind-generated electricity in China. *Science*, 2009, 325(11 September), 1378-80.

[2] Energy Enterprises Association of China. China electricity industry development year report. Beijing: China Electricity Press, 2007.

[3] Lund, H; Mathiesen, BV. The role of carbon capture and storage in a future sustainable energy system. *Energy*, 2012, 44(1), 469-76.

[4] Alberg, Østergaard, P; Mathiesen, BV; Möller, B; et al. A renewable energy scenario for Aalborg Municipality based on low-temperature geothermal heat, wind power and biomass. *Energy*, 2010, 35(12), 4892-901.

[5] Batista, NC; Melício, R; Matias, JCO; et al. Photovoltaic and wind energy systems monitoring and building/home energy management using ZigBee devices within a smart grid. *Energy*, 2013, 49(1), 306-15.

[6] Chang, X; Liu, X; Zhou, W. Hydropower in China at present and its further development. *Energy*, 2010, 35(2010), 4400-6.

[7] Zhu, E; Zhao, G. Water resources development strategies in China. Beijing: China Water Conservancy and Hydropower Press, 2002.

[8] Olabi, AG. Developments in sustainable energy and environmental protection. *Energy*, 2012, 39(1), 2-5.

[9] Olabi, AG. Developments in sustainable energy and environmental protection. *Simulation Modelling Practice and Theory*, 2011, 19(4), 1139-42.

[10] Li, GC; Huang, GH; Lin, QG; et al. Development of a GHG-mitigation oriented inexact dynamic model for regional energy system management. *Energy*, 2011, 36(5), 3388-98.

[11] Charoenngam, C; Yeh, C. Contractual risk and liability sharing in hydropower construction. *International Journal of Project Management*, 1999, 17(1), 29-37.

[12] Tang, W; Qiang, M; Duffield, CF; Young, DM; Lu, Y. Risk management in the Chinese Construction Industry. *Journal of Construction Engineering and Management*, 2007, 133(12), 944-956.

[13] Holdren, JP. Science and technology for sustainable well-being. *Science*, 2008, 319(25 January), 424-34.

[14] Vörösmarty, CJ; Mclntyre, PB; Gessner, MO; Dudgeon, D; Prusevich, A; Green, P; et al. Global threats to human water security and river biodiversity. *Nature*, 2010, 467(30), 555-61.

[15] Piao, S; Ciais, P; Huang, Y; Shen, Z; Peng, S; Li, J, et al. The impacts of climate change on water resources and agriculture in China. *Nature*, 2010, 467(2), 43-51.

[16] Rahman, MM; Kumaraswamy, MM. Joint risk management through transactionally efficient relational contracting. *Construction Management and Economics*, 2002, 20, 45-54.

[17] Tang, W, Qiang, M, Duffield, CF, Young, DM; Lu, Y. Incentives in the Chinese Construction Industry. *Journal of Construction Engineering and Management*, 2008, 134(7), 457-67.

[18] Ates, SA; Durakbasa, NM. Evaluation of corporate energy management practices of energy intensive industries in Turkey. *Energy*, 2012, 45(1), 81-91.

[19] Construction Industry Institute (CII). (1991). In search of partnering excellence, Construction Industry Development Agency, Australia.

[20] Bresnen, M; Marshall, N. Motivation, commitment and the use of incentives in partnerships and alliances. *Construction Management and Economics*, 2000, 18(5), 587-98.

[21] Bower, D; Ashby, G; Gerald, K; Smyk, W. Incentive mechanisms for project success. *Journal of Management Engineering*. 2002, 18(1), 37-43.

[22] Jaafari, A. Management of risks, uncertainties and opportunities on projects: time for a fundamental shift. *International Journal of Project Management*, 2001, 19: 89-101.

[23] Tang, W; Duffield, CF; Young, DM. Partnering mechanism in construction: an empirical study on the Chinese construction industry. *Journal of Construction Engineering and Management*, 2006, 132(3), 217-29.

[24] Tang, W; Qiang, M; Duffield, CF; Young, DM; Lu, Y. Enhancing Total Quality Management by Partnering in Construction. *Journal Professional Issues in Engineering Education and Practice*, 2009, 135(4), 129-141.

[25] Chan, APC; Chan, DWM; Fan, LCN; Lam, PTI; Yeung, JFY. Achieving partnering success through an incentive agreement: lessons learned from an underground railway extension project in Hong Kong. *Jounal of Management Engineering*, 2008, 24(3), 128-37.

[26] Love, PED; Davis, PR; Chevis, R; Edwards, DJ. Risk/reward compensation model for civil engineering infrastructure alliance projects. *Journal of Construction Engineering and Manage*, 2011, 137(2), 127-36.

[27] Eom, CSJ; Paek, JH. Risk index model for minimizing environmental disputes in construction. *Journal of Construction Engineering and Management*, 2009, 135(1), 34-41.

[28] Zou, PXW; Chen, Y; Chan, T. Understanding and improving your risk management capability: assessment model for construction organization. *Journal of Construction Engineering and Management*, 2010, 136(8), 854-863.

[29] Hoinka, K; Ziębik, A. Mathematical model for the choice of an energy management structure of complex buildings. *Energy*, 2010, 35(2), 1146-56.

[30] Love, PED; Holt, GD; Li H. Triangulation in construction management research. *Engineering Construction and Architectural Management*, 2002, 9(4), 294-303.

[31] Thomas, R. Surveys. Arnold, London: Research methods: guidance for postgraduates, 1996.

[32] Akintoye, AS; Macleod, MJ. Risk analysis and management in construction. *International Journal of Project Management*, 1997, 15(1), 31-8.

[33] SPSS Incorporated. SPSS Base 7.5 for Windows Users Guide. Chicago: Statistical Package for Social Science Incorporated, 1997.

[34] Klemes, JJ; Pistikopoulos, EN; Georgiadis, MC; Lund H. Energy systems engineering. *Energy*, 2012, 44(1), 2-5.

In: Hydroelectricity and Power Electronics
Editor: Lucille Burke
ISBN: 978-1-63321-728-7

Chapter 4

HARDWARE-IN-THE-LOOP SIMULATIONS AND CONTROL SYSTEMS BASED ON FPGA FOR POWER ELECTRONICS IN MICROGRIDS

Alonso Gutiérrez**, *Harold Chamorro,
Fernando Jiménez, Sergio Silva, Gerson Cote,
Melissa Díaz and Horacio Hoyos
Department of Electrical and Electronic Engineering,
Universidad de los Andes, Bogotá, Colombia

ABSTRACT

Nowadays, power electronics applications in microgrids have increased their complexity. This rising of complexity implies new strategies for real time simulation and control systems design. Therefore, high performance devices are required to implement Hardware-in-the-Loop models of microgrids for real time simulations. Furthermore, control system algorithms for power electronics in microgrids demand specialized hardware with optimized architecture to execute these algorithms. Taking into account the hardware requirements, FPGAs are the best hardware option to implement Hardware-in-the-Loop models and complex control systems. In order to accomplish the mentioned requirements for power electronics in microgrids, this chapter presents in detail a method to design and implement in FPGA Hardware-in-the-Loop real time simulations and control systems in microgrids. These models and control systems are described by Systems

* Corresponding author: a.gutierrez75@uniandes.edu.co.

Modeling Language (SysML). Additionally, the HiLeS-RCP (High Level Specification of embedded Systems- Rich Client Platform) methodology is employed to validate and transform the SysML diagrams in Petri nets for model implementations in FPGA. Finally, results of these implementations in embedded hardware are compared with conventional methods of simulation. To summarize, in this chapter is presented a methodology to model Hardware-in-the-Loop applications and control systems with SysML diagrams for power converters in microgrids; in addition, the methodology proposed transforms these SysML diagrams to Petri nets for their validation and implementation in FPGA.

1. INTRODUCTION

Power electronics is the technology of applying solid-state electronics for the control and conversion of electric power [1]. Power electronics in microgrids considers several applications [2] such as power converters to control the power flow of solar photovoltaic with the electrical grid, inverters to interface variable speed wind turbines and the grid, power control devices to integrated electric vehicles into the grid, and power devices to enable power grid dynamic stabilization. In addition, the increased penetration of microgrids featuring higher flexibility and reliability shows new challenges for conventional power electronics systems [3]. As a consequence, the power electronics growth and the integration of alternative energy resources require power electronics equipment with more efficiency and flexibility [4]. Furthermore, power electronics systems with innovative control strategies are needed to accomplish the current requirements of power electronics [5]. Nonetheless, power electronics in microgrids requires more powerful processing devices because of the increment of its control complexity [6]. According to [7], the design of complex control systems for power electronics also involves real-time emulation tools. In fact, Hardware-in-the-Loop (HIL) techniques are proposed to design and implement complex control systems in power electronics for microgrids.

Control systems for power electronics in microgrids require high performance devices; as a consequence, solutions based on FPGA arise as suitable alternatives. However, these control system implementation in FPGA also require novel schemes to accomplish the control specifications [8]. Therefore, new methodologies are required to increase the productivity and to decrease the implementation in hardware efforts. Taking into account the previous factors, HiLeS-RCP (High Level Specification of Embedded Systems

- Rich Client Platform) [9] is proposed as a design methodology for HIL in power electronics and control systems based on FPGA in microgrids.

HiLeS-RCP[10] is a tool and methodology intended to generate Petri net representations and VHDL-AMS code from SysML models. HiLeS-RCP platform is developed to design and implement complex electronic systems in FPGA. These complex electronic systems include control systems and Hardware-in-the-Loop models represented by SysML diagrams. Moreover, the HiLeS-RCP characteristic of converting SysML diagrams in Petri nets[11], which allows validating the SysML models from Petri net representations. Additionally, structural validation of SysML models through Petri nets allows guaranteeing the structural stability previous to implementation in embedded hardware. Consequently, HiLeS-RCP is shown in this chapter as a suitable tool and methodology to aid designers in the process of system implementation in FPGA; in this specific case, to implement the controllers of interleaved boost converters.

In the following sections, first are presented Hardware-in-the-Loop models for simulation and control design of microgrids components; these models and control systems are described by Systems Modeling Language (SysML) because SysML supports the specification, analysis, design, verification and validation of a wide diversity of systems. Next, the HiLeS-RCP (High Level Specification of embedded Systems- Rich Client Platform) methodology is employed to validate and transform these SysML diagrams to Petri nets; the Petri nets offer a graphical representation for discrete event systems with concurrent execution ideal for implementations in FPGA. Then, the Petri nets from SysML diagrams of microgrid components are modeled in VHDL for implementation, and the implementation results are analyzed. Finally, results of these implementations and simulations are compared with conventional methods of simulation.

2. Hardware-in-the-Loop for Microgrid Simulations

Hardware-in-the-Loop (HIL) is a technique used to develop complex real-time systems. HIL simulations provide a suitable platform to assess complex systems and control actions. The complexity of these systems and their control actions can be included in the simulation processes by means of mathematical representation of dynamic system models[12]. HIL simulations include electrical emulation of sensors and actuators which act as interface between the simulated system and the embedded control system. Then, the emulated sensors are controlled by the simulated system, and the control algorithms send control signals to the emulated actuators. As a result, changes in the control signals result in changes in the simulated system. The use of Hardware-in-the-Loop simulations allows the design and implementation of control systems in embedded hardware. Indeed, embedded control system could be tested in emulated systems without most of the real plant limitations. For instance, testing a power electronics control as a real plant can create dangerous conditions for the power device.

Power electronics simulations in microgrids are currently driven to Hardware-in-the-Loop technology. This technology allows the design and power electronics test of microgrid systems reducing the development cycle, increasing efficiency, and improving reliability and safety of microgrid systems. In fact, power electronics are systems with fast dynamics because of high-frequency switching action. As a consequence, real-time simulations of switching transitions require digital processor that can be met by FPGA technologies achievement high performance simulations of power electronics in microgrids[13].

HIL simulations for microgrid systems also are used to verify the stability, operation, and fault tolerance of microgrids[14]. Real-time emulations in microgrids have the ability to model power systems that includes several components associated to generators, loads, alternative energy sources, and electrical grid interconnections. These types of simulation allow the assessment and analysis in real-time microgrid power systems in realistic emulated environment. Furthermore, Hardware-in-the-Loop (HIL) in microgrids has been used for exploring the integration of renewable energy resources, and innovative control systems.

3. HiLeS-RCP for Hardware-in-the-Loop Simulations

Hardware-in-the-Loop simulations base on embedded systems require tools for helping to designers to decrease the time-to-market, establish economic viability, and build complete virtual prototypes. Nevertheless, HiLeS-RCP provided a tool for designers to produce high level descriptions of embedded systems defining the functional composition, hierarchy and structure. HiLeS-RCP platform is based on SysML and Petri Nets to formal design and verification[15]. In addition, HiLeS-RCP generates partial VHDL-AMS code from SysML models and provides functionality validation by the evaluation of Petri nets using TINA(TIme petri Net Analyser)[16]. As a consequence, HiLeS-RCP is a suitable platform to develop Hardware-in-the-Loop simulations based on embedded hardware.

3.1. HiLeS Formalism

Petri net is a formal language which is used to model discrete events systems[17]. Petri nets are based on graphical notation to represent sequential and concurrent process. Petri nets are currently used to model and verify designs in different domains. HiLeS (High Level Specification of Embedded Systems) formalism[18] consists of a language and a methodology based on Petri nets which incorporates continuous behavior and hierarchy. The HiLeS formalism is intended to support embedded system design. Furthermore, HiLeS formalism defines the rules to represent a HiLeS model in hardware description languages such as VHDL-AMS. These characteristics enable designers to build virtual prototypes allowing technological independence, operational verifications, and formal verification of specifications.

The HiLeS formalism is intended to support the processes of designing formal behavioral models and to integrate features of high level graphic semantic. HiLeS allows describing asynchronous and concurrent behaviors; also, it describes time-driven event or mixed-driven dynamics. The semantics of the HiLeS formalism is presented as a HiLeS Control Net (HCN) which is an extension of Petri nets. The HiLeS Control Net (HCN) extends the Petri Net definition by including Structural and Functional Blocks (the structural elements of the HiLeS formalism). A Marked Petri Net (MPN) is a 5-tuple [19]:

$$MPN = \langle P, T, I, O, M \rangle \quad (1)$$

where:

$P = \{p_1, p_2, \dots, p_n\}$ is a finite set of places;
$T = \{t_1, t_2, \dots, t_m\}$ is a finite set of transitions;
$I \subseteq (P \times T) \cup (T \times P)$ is a set arcs;
O: F where N is a weight function;

M: *P* where *N* is a marking function, M_0 is the initial marking;
$P \cap T = \phi$ *and* $P \cup T \neq \phi$

The semantics of a PN is defined by the firing of enabled transitions, and can be described as follows [19]:

- A transition t is enabled if each input place p is marked with at least o(p,t) tokens, where o(p,t) is the weight of the arc from p to t.
- An enabled transition may or may not fire; depending on wheatear the event actually takes place.
- A firing of an enabled transition t removes o(p,t) tokens from each input place p of t and adds o(p,t) tokens to each output place p of t.

An HCN (HiLeS Control Net) can be expressed as a 5-tuple [19]

$$HCN = \langle MPN, \boldsymbol{FB}, \boldsymbol{SB}, \boldsymbol{J}, R \rangle \qquad (2)$$

Where:

MPN is a marked PN;
$FB = \{f_1, f_2, \dots, f_n\}$ *is a finite set of Functional blocks;*
$SB = \{s_1, s_2, \dots, s_n\}$ *is a finite set of Structural blocks;*
$J \subseteq (T \times SB) \cup (SB \times T) \cup (P \times SB) \cup (SB \times P) \cup (T \times FB) \cup (FB \times T)$ *is a set arcs;*
R: F where N is a weight function;
$SB \cap FB = \phi$ *and* $SB \cup FB \neq \phi$*;*
Weight of all arcs must be 1, i.e., the HCN is ordinary.

According to [19], the two reasons for extending the definition of PN are:

- To provide the HiLeS Control Net (HCN) with the hierarchy do not defined in Petri nets.
- To use the signaling mechanism of Petri nets (tokens) to represent the execution of system components.

The behavior of the Petri nets extended by the HCN definition because of the use of Structural Blocks and Functional Blocks elements. Thus, the semantics of a HCN can be expressed as follows [19]:

- A transition *t* is enabled if each input place *p*, each input Structural Block *SB*, and each input Functional Block *FB*, associated to this transitions *t*, is marked by at least one token.
- An enabled transition *t* may be fired if the event actually takes place.
- A firing of an enabled transition *t* removes one token from each input place *p*, each input Structural Block *SB*, and each input Functional Block *FB*, associated to this transition *t*. Adding one token to each output place *p*, each output Structural Block *SB*, and each output Functional Block *FB* next to this transition *t*.

3.2. SysML Models in HiLeS-RCP

SysML (Systems Modeling Language) is an international standard language for modeling systems and systems of systems[20]. SysML supports models based on requirement definition, analysis, and validation of a wide variety of systems in engineering fields [21]. Representative diagrams of SysML are Block Definition Diagrams, Internal Block Diagrams, and Activity Diagrams.

SysML uses Block Definition Diagrams (BDD) to describe structural composition of systems and Internal Block Diagram (IBD) to define the interconnections among the system components. Figure 1 shows the partial Block Definition Diagram (BDD) and the Internal Block Diagram (IBD) of an interleaved power converter. In Figure 1 the interleaved power converter is built by two boost converters (Figure 1a); additionally, the interconnections of these two converters are represented by the Figure 1b.

SysML uses BDD and IBD to define the structure of a system. Furthermore, SysML describe the system behavior by means of Activity Diagrams. Figure 2 shows the partial Activity Diagram of a control system for an interleaved power converter. Figure 2 represents the sequence of the control system and the data flow.

HiLeS-RCP is a platform intended to generate Petri net representations and VHDL code from SysML models [22]. Figure 3 shows the SysML transformation chain used by HiLeS-RCP. In Figure 3, transformation T1 provides the synthesis from SysML models to HiLeS formalism. T2 is used to produce Petri nets representations from HiLeS models. Additionally, transformation T2 transforms the Petri net in textual format to validate the Petri nets by TIme petri Net Analyser (TINA). Finally, transformation T3 is used to transform HiLeS description in VHDL language.

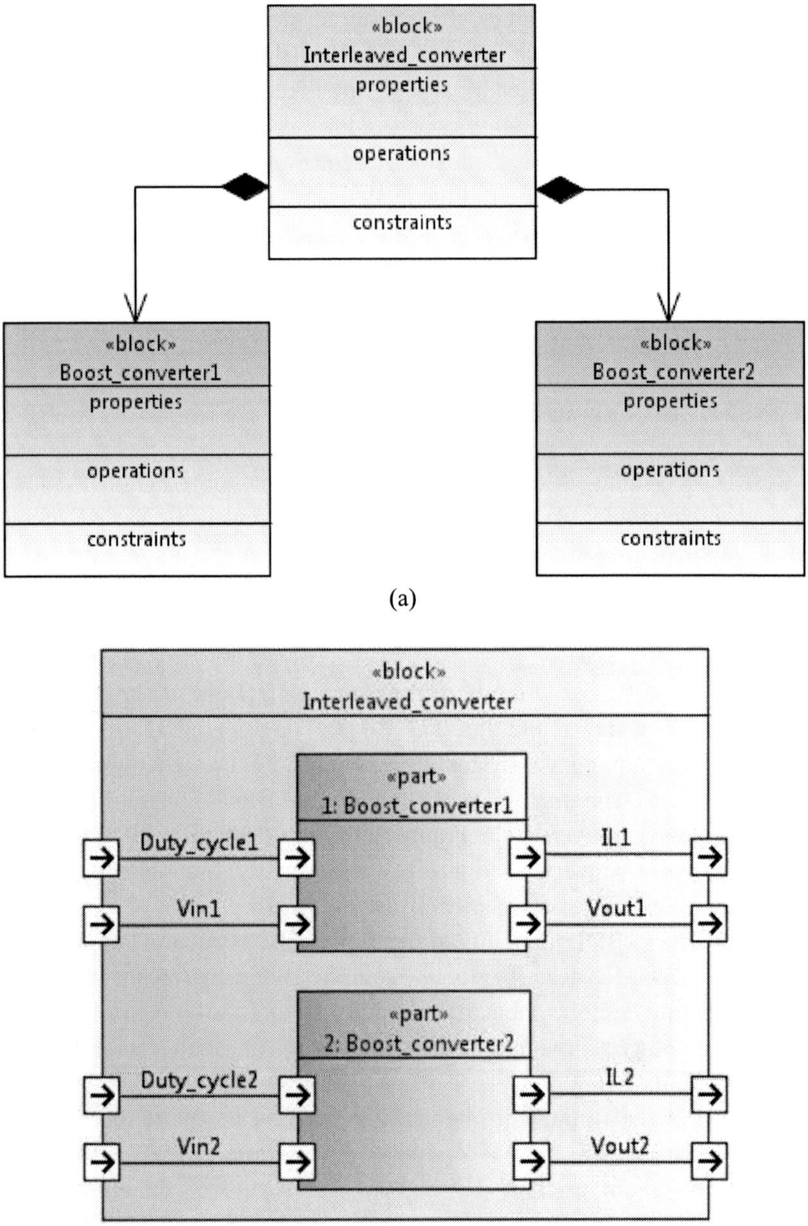

(b)

Figure 1. Partial Block Definition Diagram (a) and Internal Block Diagram (b) of a power converter.

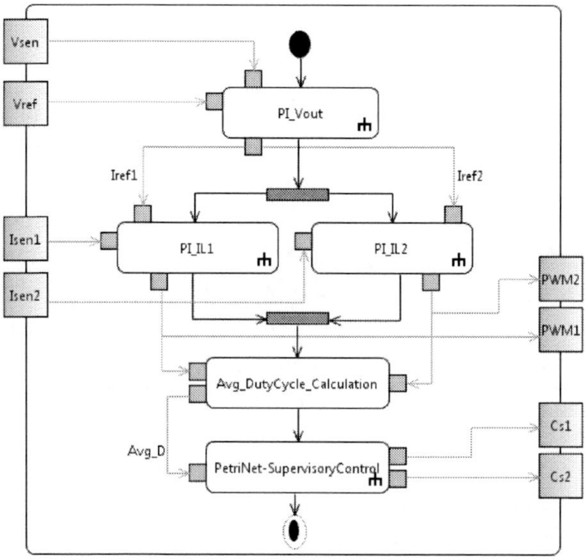

Figure 2. Partial Activity Diagram of a control system for an interleaved power converter.

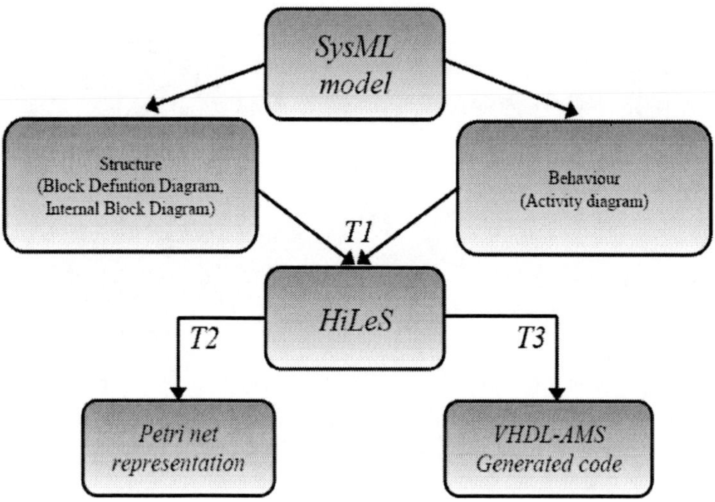

Figure 3. SysML transformation chain in HiLeS-RCP.

3.3. Behavior Transformation

The behavior transformation consists of creating the required HiLeS *Control Net* (HCN) from SysML Activity Diagrams (Figure 4). The main advantages of using Activity Diagrams in SysML models are their direct representation in Petri nets. A detailed description of this transformation is found in [10]. In this approach, there are two types of actions described in Activity Diagrams: Call BehaviorActions and Call Operation Actions. *CallBehaviorAction* are used to invoke other Activity Diagrams equivalent to hierarchical construction. As result, CallBehaviorActions creates structural block and adds the required HCN elements to correctly represent the model behavior. In addition, the CallOperationAction are transformed in the same way, but instead of a structural block a functional block is created.

The transformation from SysML models to HiLeS Control Net based on Petri nets allows the automatic VHDL code generation. The following script represents a partial VHDL code generation from a HiLeS Control Net (HCN).

The VHDL code generated using HiLeS-RCP can be synthesizable in FPGA as shown in Figure 6.

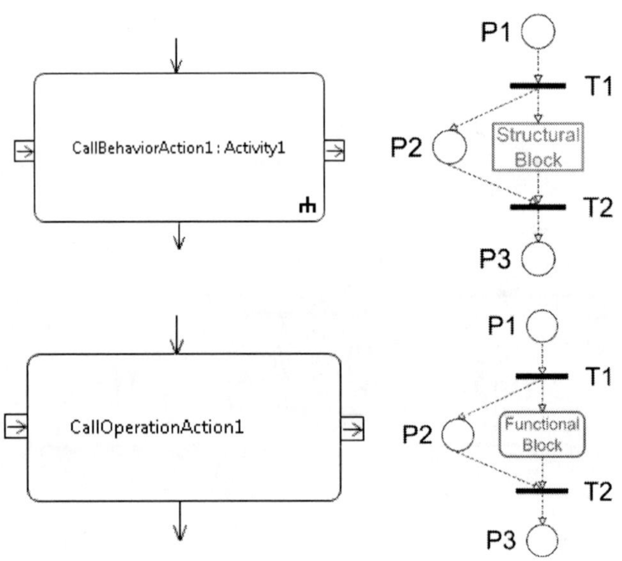

Figure 4. HiLeS *Control Net* (HCN) from SysML Activity Diagrams.

```
entity Place_Sync is
generic(  nb_entrees   : natural:= 1;
          nb_sorties   : natural:= 1);
port  (clk,init    : in std_logic;
       ini_jetons    : in std_logic;
       aj_jetons    : in std_logic_vector(nb_entrees-1 downto 0);
       ret_jetons    : in std_logic_vector(nb_sorties-1 downto 0);
       marque    : out std_logic);
end Place_Sync;

architecture Behavioral of Place_Sync is
   signal D, X, Q : std_logic;
   signal ret, aj : std_logic;
begin
   process (ret_jetons, ini_jetons, init)
          variable pl : std_logic;
          variable Inic : std_logic;
   begin
          pl := ret_jetons(0);
          ciclo_r_and: for i in 1 to nb_sorties-1 loop
                 pl := pl and ret_jetons(i);
          end loop;
          ret <= pl;
   end process;
   process (aj_jetons)
          variable tl : std_logic;
          begin
          tl := aj_jetons(0);
          ciclo_a_and: for i in 1 to nb_entrees-1 loop
                 tl := tl and aj_jetons(i);
          end loop;
           aj<= tl;
          end process;
          D<= aj and (not ret);
          X<=aj or ret;
   process(clk)
begin
  if rising_edge(clk) then
    if init = '1' then
      Q <= ini_jetons;

          elsif X<='0' then
      Q <= Q;
          else
          Q<= D;
    end if;
  end if;
end process;
   marque<=Q;
end Behavioral;
```

Figure 5. Partial script of VHDL code generated by HiLeS-RCP.

3.4. SysML and HIL Models in Microgrids

The use of HiLeS-RCP allows the hardware implementation of SysML diagrams. These SysML diagrams can describe complex systems of Hardware-in-the-Loop models to applications in power electronics and control systems in microgrids. Modeling microgrids with SysML and HIL increases the simulation performance and allows their implementation in embedded hardware when tools such as HiLeS-RCP are employed.

4. HARDWARE-IN-THE-LOOP MODEL OF MICROGRID

The Hardware-in-the-Loop simulations are useful to verification complex systems such as microgrids. Figure 7 presents an example of a Microgrid built by a photovoltaic system with parallel DC-DC converters and a grid connected inverter. The control design of this system, together with the Maximum Power Point Tracking algorithm can be tested on FPGA using Hardware-in-the-Loop before the practical implementation. In this section is presented the Hardware-in-the-Loop model of microgrid shown in Figure 7, in addition the components of the proposed microgrid are modeled by SysML and HiLeS-RCP.

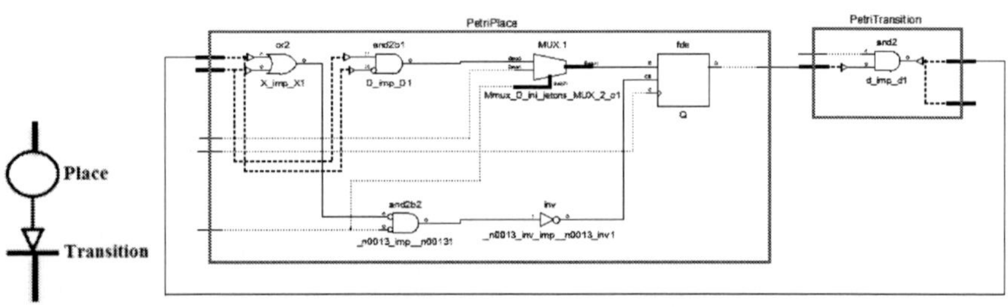

Figure 6. Synthesis of VHDL code generated by HiLeS-RCP.

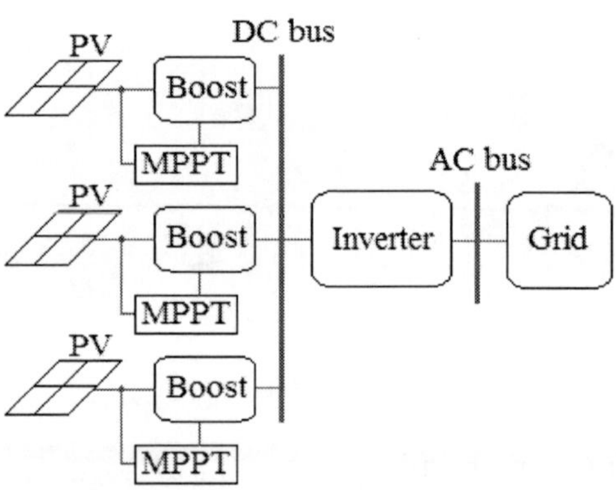

Figure 7. Example of a Microgrid.

4.1. Boost Converter Model

Boost converters are power converters of wide application in photovoltaic systems. In a boost converter the input voltage is increased; as a consequence, the output voltage is higher than the input voltage[23]. Figure 8 shows a conventional boost converter circuit including losses associated to inductor (R_L). This simplification is suitable to design the Hardware-in-the-Loop model without loss of accuracy.

The behavior of a boost converter is described in Figure 8. According to [23], when PWM signal is high, the switch is ON. As a result, the inductor voltage and the capacitor current when the switch is in state ON is given by,

$$v_L = V_{in} - i_L R_1 \tag{3}$$

$$i_c = -\frac{v_C}{R} \tag{4}$$

The inductor voltage and the capacitor current during the switch state OFF is given by,

$$v_L = V_{in} - i_L R_1 - v_C \tag{5}$$

$$i_c = i_L - \frac{v_C}{R} \tag{6}$$

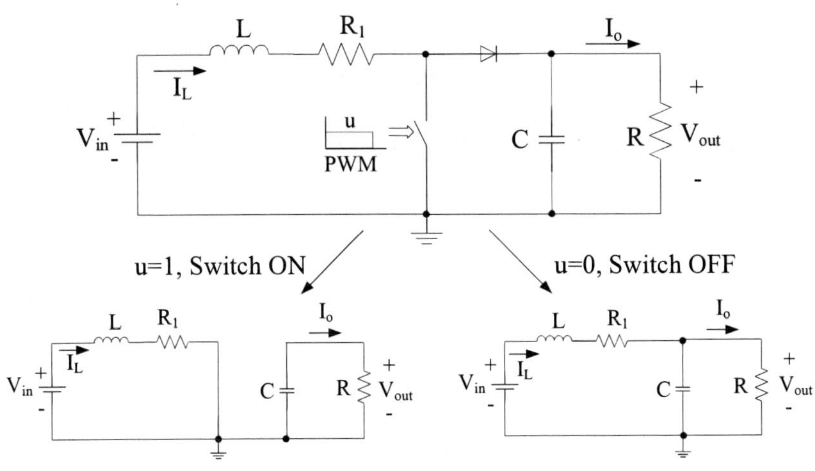

Figure 8. Boost converter.

Given the control signal $= \{1\, for\ state\ ON, 0\, for\ state\ OFF\}$,

$$v_L = V_{in} - i_L R_1 - v_c(1-u) \tag{7}$$

$$i_c = i_L(1-u) - \frac{v_c}{R} \tag{8}$$

In terms of differential equations,

$$C\frac{dv_c}{dt} = i_L(1-u) - \frac{v_c}{R} \tag{9}$$

$$L\frac{di_L}{dt} = V_{in} - i_L R_1 - v_c(1-u) \tag{10}$$

Given the discrete form with a sample time h,

$$C\left[\frac{v_c(k+1) - v_c(k)}{h}\right] = i_L(1-u(k)) - \frac{v_c(k)}{R} \tag{11}$$

$$L\left[\frac{i_L(k+1) - i_L(k)}{h}\right] = V_{in} - i_L R_1 - v_c(1-u) \tag{12}$$

Solving for $i_L(k+1)$ and $v_c(k+1)$,

$$v_c(k+1) = \frac{h}{C}\left[i_L(1-u) - \frac{v_c}{R}\right] + v_c(k) \tag{13}$$

$$i_L(k+1) = \frac{h}{L}[V_{in} - i_L(k)R_1 - v_c(k)(1-u)] + i_L(k) \tag{14}$$

The output voltage and current are given by,

$$v_{out}(k) = v_c(k) \tag{15}$$

The load model is defined by,

$$i_o(k) = \frac{v_{out}(k)}{R} \tag{16}$$

Defining K_1 and K_2,

$$K_1 = \frac{h}{C} \tag{17}$$

$$K_2 = \frac{h}{L} \tag{18}$$

Thus,

$$v_c(k+1) = K_1\left[i_L(1-u) - i_o(k)\right] + v_c(k) \tag{19}$$

$$i_L(k+1) = K_2\left[V_{in} - i_L(k)R_1 - v_c(k)(1-u(k))\right] + i_L(k) \tag{20}$$

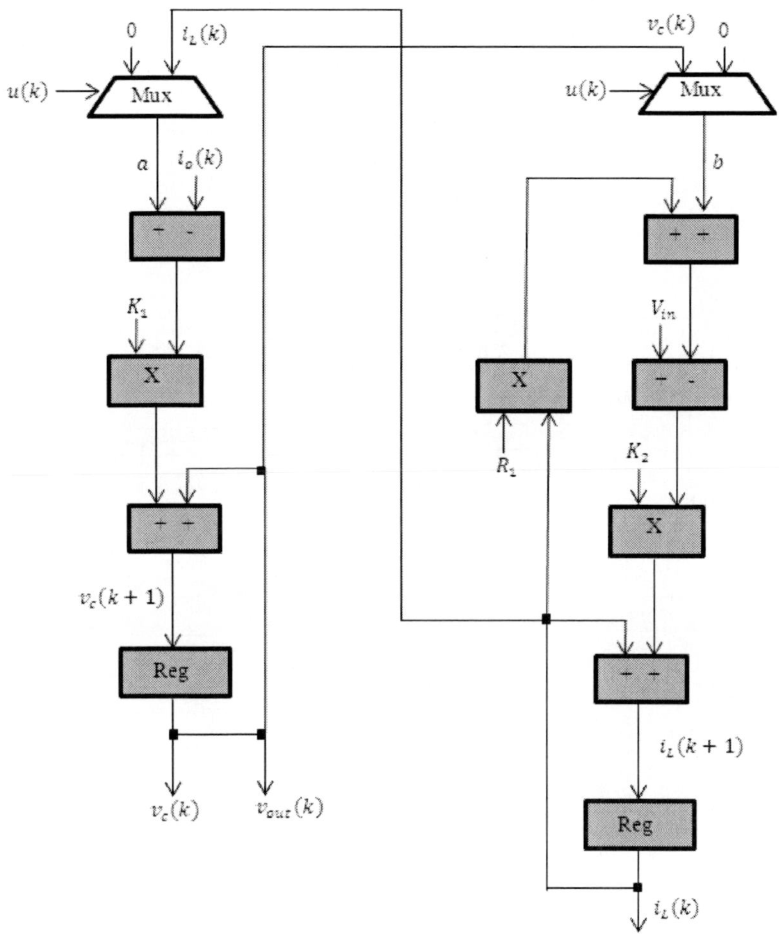

Figure 9. Data flow graph to the boost converter model.

Defining a and b,

$$a = i_L(k)(1 - u(k)) = \begin{cases} 0, u(k) = 1 \\ i_L(k), u(k) = 0 \end{cases} \tag{21}$$

$$b = v_c(k)(1 - u(k)) = \begin{cases} 0, u(k) = 1 \\ v_c(k), u(k) = 0 \end{cases} \tag{22}$$

The variables a and b can be represent as a multiplexer; then, the next figure shows the data flow graph corresponding to the boost converter model.

The power system for the boost converter modeled by SysML diagrams is described in figures 10, 11, and 12.

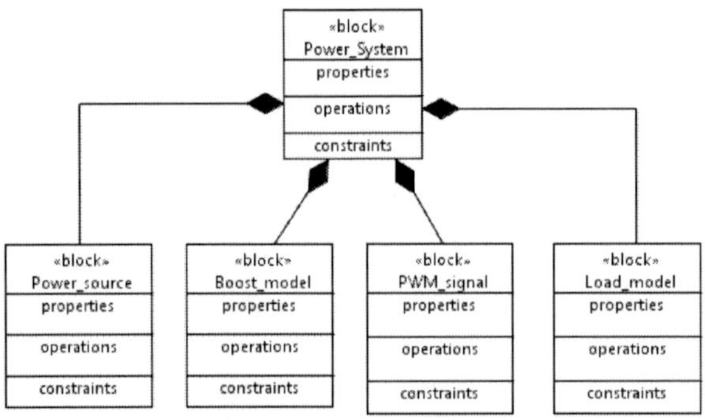

Figure 10. BDD of boost converter system.

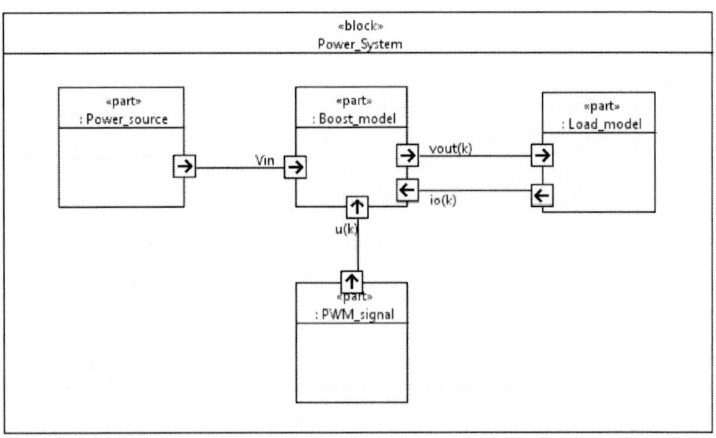

Figure 11. IBD of boost converter system.

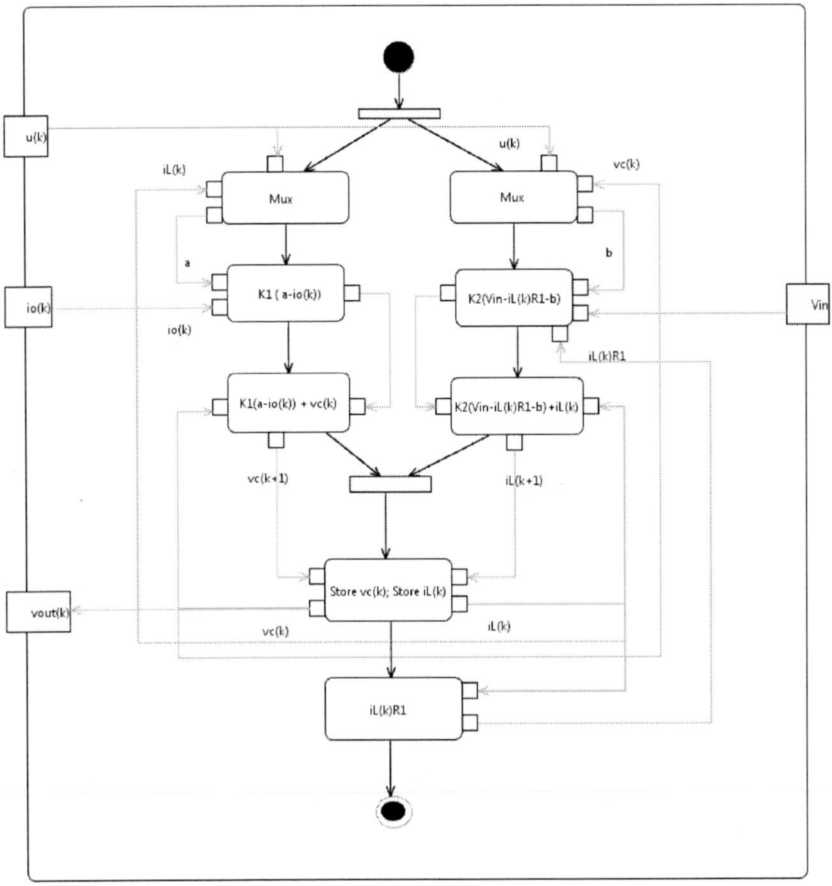

Figure 12. Activity diagram of boost converter system.

The asociated Control Petri Nets from the SysML model of the boost converter system in shown in Figure 13.

The implementation in embedded hardware of the boost converter system modeled by SysML and HiLeS-RCP is presented in the following figures. Math operations in FPGA are calculated in float point format with simple precision (IEEE 754 standard). The result validation was made in the software Systen Vision.

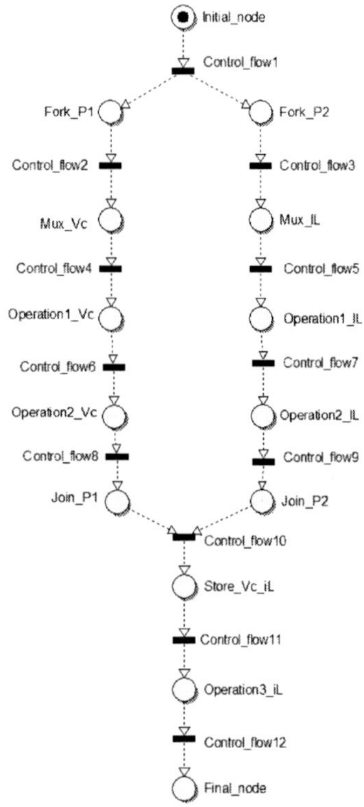

Figure 13. Control Petri net from boost converter model.

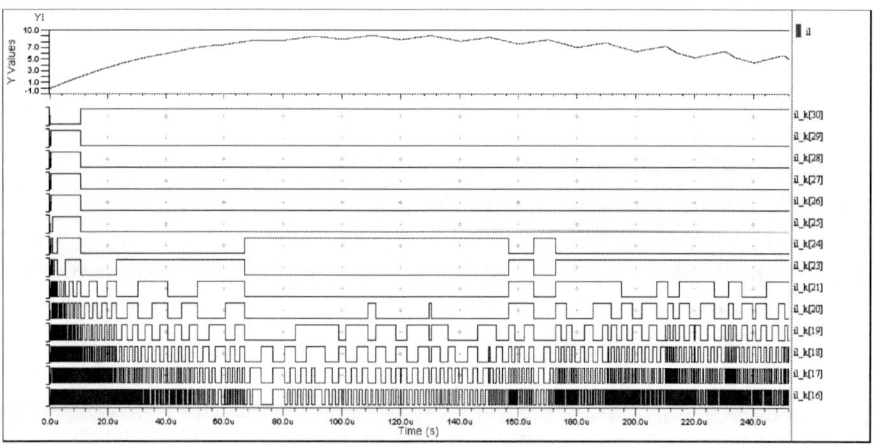

Figure 14. Boost converter model in VHDL.

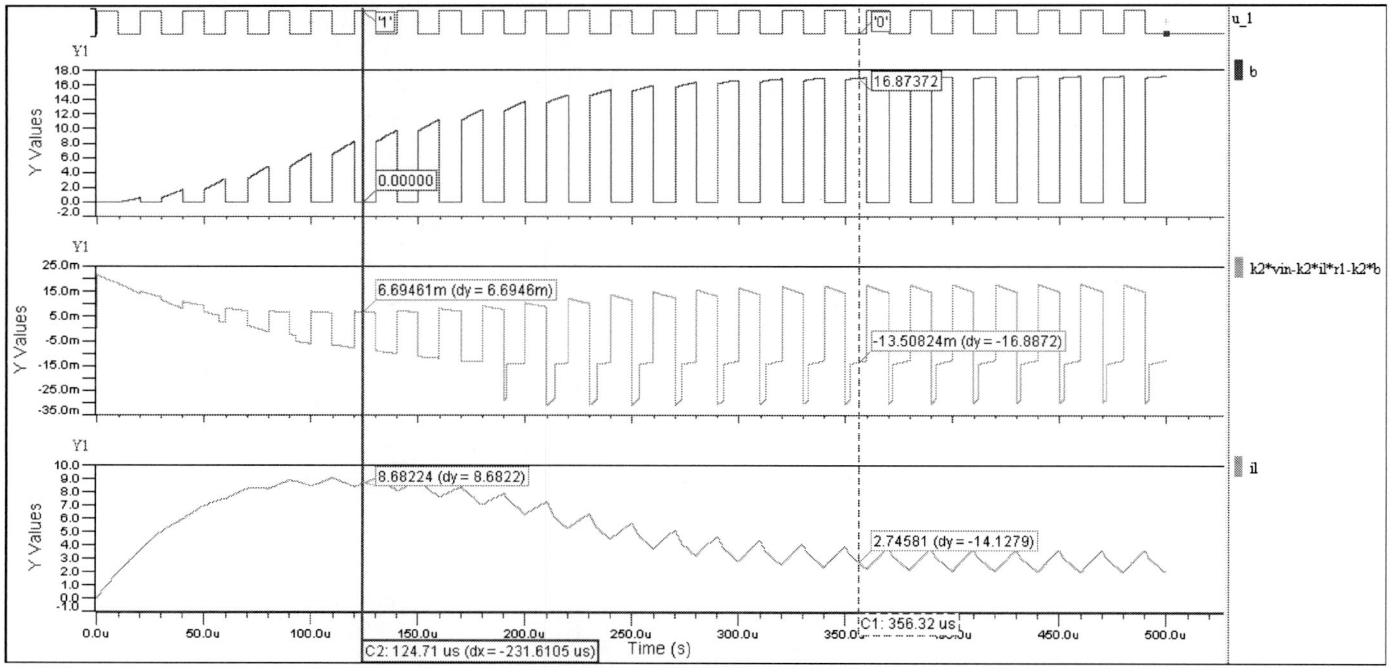

Figure 15. Inductor curruent calculation of boost converter model using VHDL.

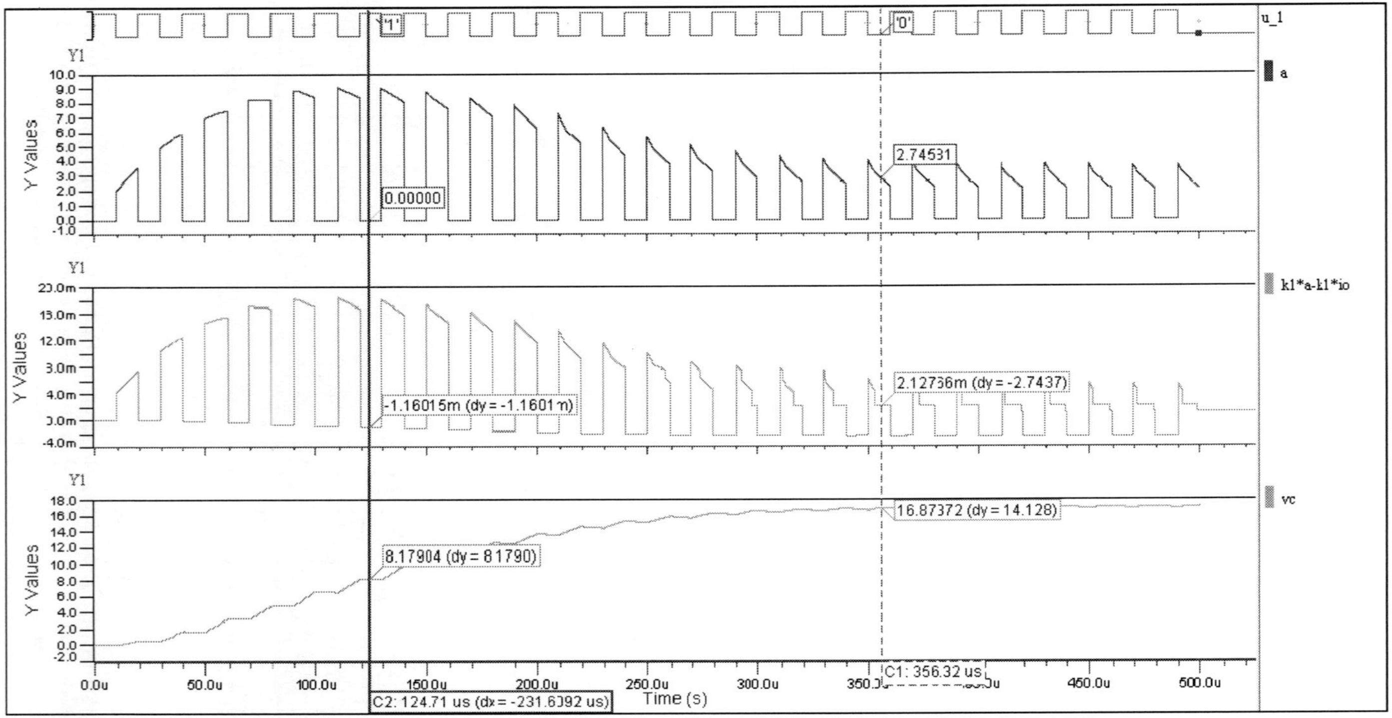

Figure 16. Capacitor voltage calculation of boost converter model using VHDL.

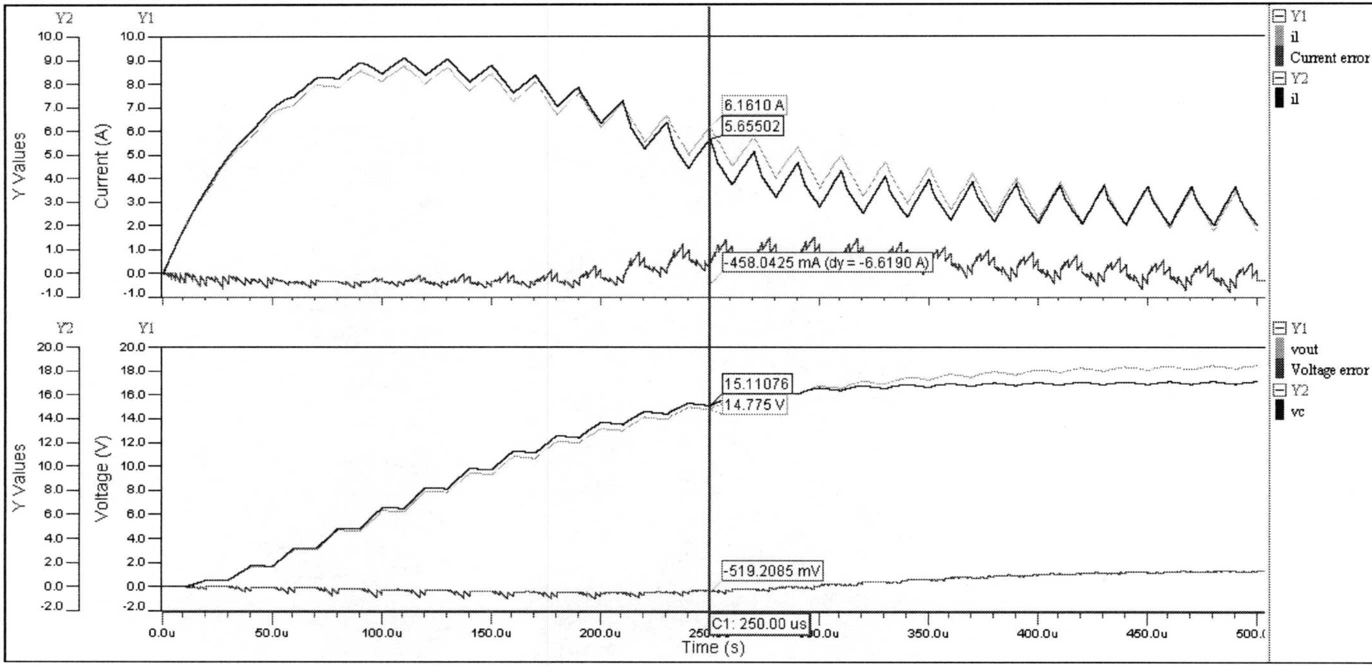

Figure 17. Comparison of boost converter model using VHDL with conventional circuit simulation.

The hardware implementation is made using FPGA and high speed DAC to display the current and voltage signals.

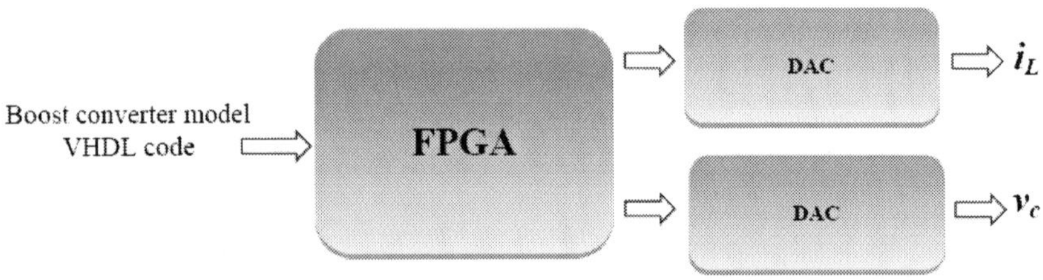

Figure 18. Implementation in hardware.

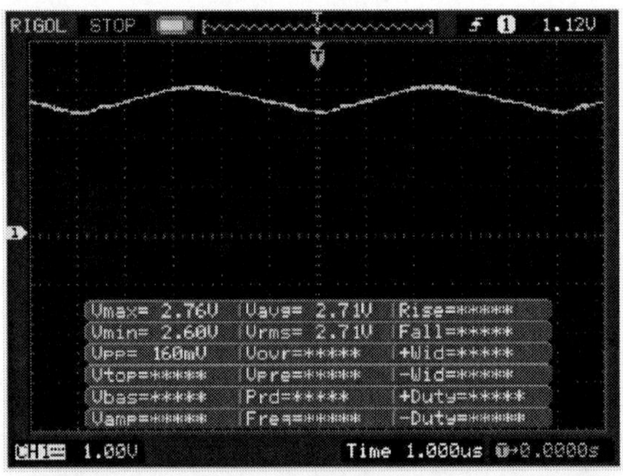

(a)

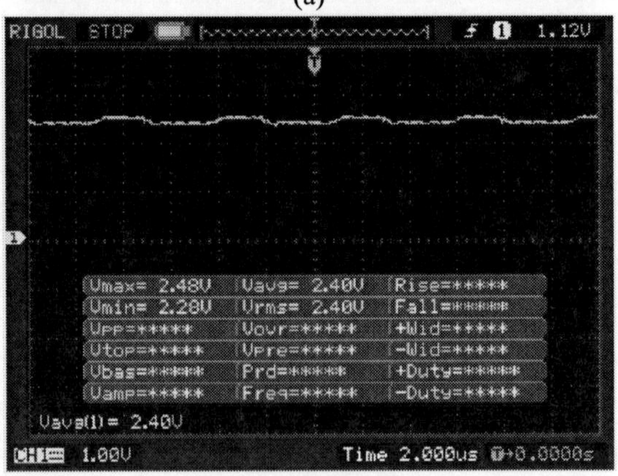

(b)

Figure 19. a) inductor current, b) capacitor voltage.

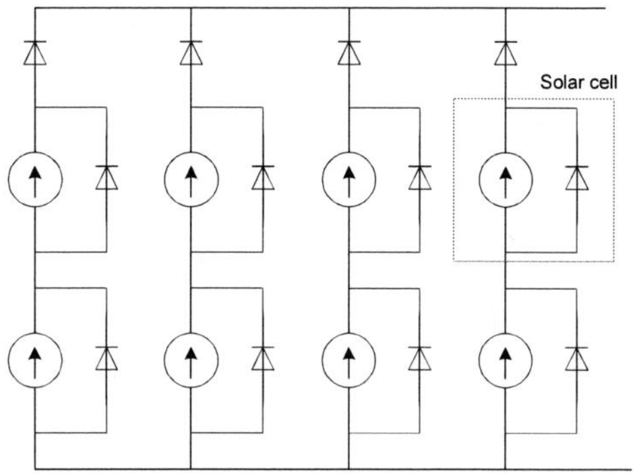

Figure 20. Solar cells array conforming a photovoltaic module [24].

4.2. Photovoltaic Module

The photovoltaic module is a device conformed by solar cells arrays. Those cells are semiconductor unions (PN unions) which work with the photoelectric effect. When the photons interact with the PN union, it is created an electric potential that turns the sun energy into electric current. Each one of these cells is capable to create a potential and generate an electric current. This potential grows in the photovoltaic module when there are connected solar cells in series. This current grows when there are connected solar cells in parallel. In Figure 20 is shown an array of cells connected in parallel and in series forming a photovoltaic module.

Each photovoltaic module has its own characteristic current-voltage behavior; this is known as the characteristic I-V curve. This curve is given by the material characteristics, how is conformed the cells array, environmental parameters, as the temperature of operation and incident solar radiation, and circuit parameters as the devices connected to the photovoltaic module.

4.2.1. Photovoltaic model

There are several photovoltaic modules (PVM) models, many of these use circuital parameters that are not given by the manufacturer such as internal resistances and capacitances. The model developed in application is capable of reproduce the characteristic I-V curve using only the parameters that appears in the photovoltaic modules' datasheets, this model is restricted to modules composed by semiconductors. The model defines de current over the PVM in function of environmental parameters and operation conditions.

Expression (23) shows the current as function of voltage, a nominal current (I_x) and a nominal voltage (V_x) given by the characteristics of the panel and environmental parameters. Furthermore, the current is function of a constant "b", this constant is and adjustment parameter calculated with parameters that appear in datasheets. There is possible to find and expression for the power given by the panel using the expression (23) and multiplying it with the voltage, expression (24) result of this operation. The differential equations of current and power shown in (25) and (26) are derived of (23) and (24) respectively in function of the voltage [25].

$$I(V) = \frac{I_x}{1-e^{-\frac{1}{b}}} * \left(1 - e^{\frac{V}{b*V_x}-\frac{1}{b}}\right) \tag{23}$$

$$P(V) = V * I = \frac{V*I_x}{1-e^{-\frac{1}{b}}} * \left(1 - e^{\frac{V}{b*V_x}-\frac{1}{b}}\right) \tag{24}$$

$$\frac{dI(V)}{dV} = -\frac{I_x*e^{\frac{V}{b*V_x}}}{\left(e^{\frac{1}{b}}-1\right)*b*V_x} \tag{25}$$

$$\frac{dP(V)}{dV} = -\frac{I_x*\left((V+b*V_x)e^{\frac{V}{b*V_x}}-e^{\frac{1}{b}}*b*V_x\right)}{\left(e^{\frac{1}{b}}-1\right)*b*V_x} \tag{26}$$

Where,

$$V_x = s * \left(\frac{E_i}{E_{in}}\right) * TCv * (T - T_N) + s * V_{max} - s * (V_{max} - V_{min}) * e^{\frac{E_i}{E_{in}}*\ln\left(\frac{V_{max}-V_{oc}}{V_{max}-V_{min}}\right)} \tag{27}$$

$$I_x = p * \frac{E_i}{E_{in}} * \left(I_{sc} + TCi * (T - T_N)\right) \tag{28}$$

Parameters definition:

V	Out voltage [V]
I	Out current [A]
P	Power given by the PVM [W]

V_{oc}	Open circuit voltage (V)
I_{sc}	Short circuit current (A)
V_{mppt}	MPP Voltage (V)
I_{mppt}	MPP Current (A)
E_i	Incident intensity $[\frac{W}{m^2}]$
E_{in}	Intensity STC $\left(1000\frac{w}{m^2}\right)$
T	PVM Temperature [°C]
T_N	Temperature STC (25 °C)
TCv	Voltage sensitivity [V/°C]
TCi	Current sensitivity [A/°C]
B	Characteristic constant
s	Series PVMs
p	Parallel PVMs

Meanwhile, the constants V_{max} *and* V_{min} are usually $V_{max} = V_{oc} * 1.03$ and $V_{min} = V_{oc} * 0.85$.

4.2.2. Calculation of the characteristic constant "b"

This model requires an adjustment constant "b", called the characteristic I-V curve. This constant is related with the fill factor. This one describes the behavior of the current and of the power in the PVM. The power given by the PVM can be represented in a I-V figure by the rectangle formed when one point in the figure is selected. The fill factor is an indicator that shows the percentage of the total I-V curve area fill by this rectangle. The "b" constant is calculated with an iterative algorithm as there explained in Figure 21 and shown next:

Table 1. Parameters of the PVM given by the manufacturer, chosen for validate the PVM model

Reference	Voc (V)	Isc (A)	Vmp (V)	Imp (A)	TCv (V/°C)	TCi (A/°C)
MSX-5	20,5	0,3	16,5	0,27	-0,08	0,0002
MSX-10	21	0,65	16,8	0,59	-0,08	0,0002
SLK60M6L	37,2	7,52	30,6	6,86	-0,127	0,0022

$$Ab_0 = Isc, \quad b_0 = 0$$

$$While \ |b_{n+1} - b_n| < e_0$$

$\{$

$$b_{n+1} = \frac{\frac{V_{mppt}}{V_{oc}} - 1}{\ln 1 - \frac{I_{mppt}}{Ab_n}}$$

$$Ab_{n+1} = \frac{Isc}{1 - e^{-\frac{1}{b}}}$$

$\}$

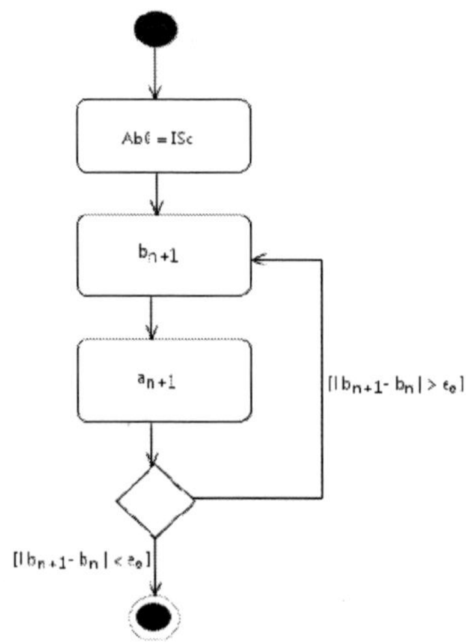

Figure 21. Activity Diagram for the characteristic constant calculation.

4.2.3. Validation of the model

The PVM presents sensitivity to the temperature and the radiation. With the objective to validate this model, there will be proved its sensitivity to temperature and to the solar radiation separately. There are used the panel references listed in [26] and [27]. There are shown in TABLE I the parameters given by the datasheet, the same all that are used in the model. Below are the sensitivity to temperature corroborated with panels MSX5 and MSX-10 and the sensitivity to radiation corroborated with panel SLK60M6L.

4.2.4. Sensitivity to radiation

The radiation in the panel varies according to the latitude of the place, the day of the year, the hour of the day, the inclination of the panel, etc. The model selected use the effective solar radiation, there are plenty ways to calculate this radiation as function of the solar radiation. Figure 5 shows the I-V curve and the P-v curve calculated with the model for intensities of 1000, 800, 600, 400 and 200 $\frac{W}{m^2}$. Figure 6 shows the real I-V and P-V curves with the same radiations listed above.

As the effective solar radiation over the PVM increases, the current delivered by the PVM and the Voc of this one is greater. The power of the PVM increases with the effective radiation also. TABLE II shows that error between the model and the values given by the manufacturer is less than 1.5% for short circuit current and open circuit voltage.

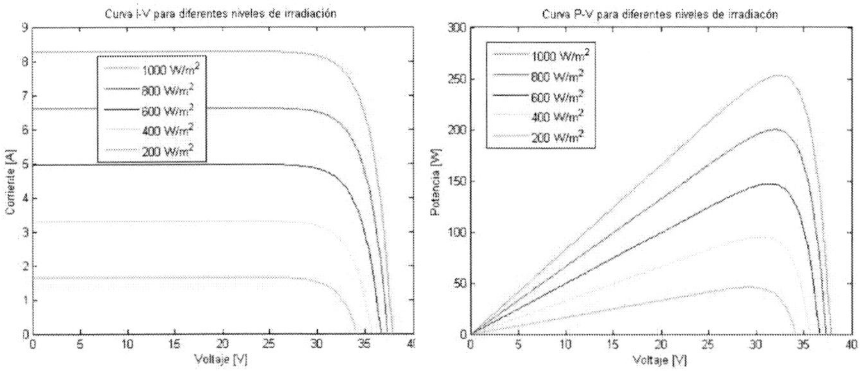

Figure 22. Simulation of I-V and P-V curves of the panel SLK60M6L varying the solar radiation.

The model presented above is able to reproduce the behavior of the PVM reproducing the I-V and the P-V curves. The model presents an adequate modeling for the sensitivity to temperature and radiation calculating values with a small absolute error in comparison to the real values given by the manufacturer.

4.2.5. Photovoltaic module in SysML

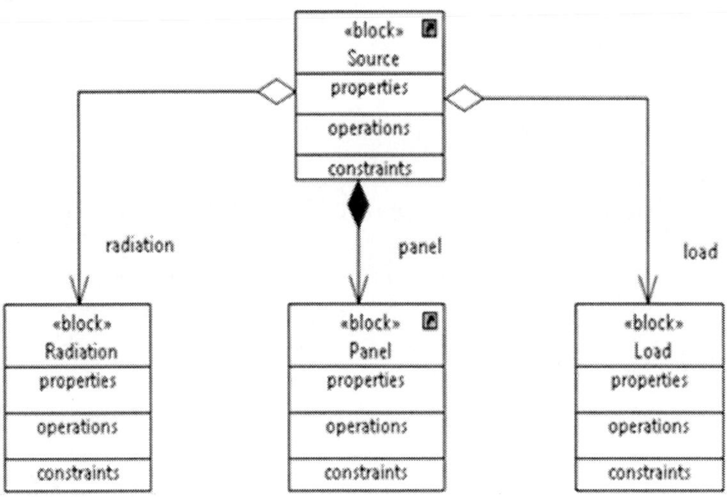

Figure 23. BDD of PV model.

Table 2. Comparison between the model and the real values given by the manufacturer for radiation variation

Ein (W/m ^2)	Voc model (V)	Isc model (A)	Voc PVM (V)	Isc PVM (A)	Error Voc	Error Isc
1000	37,5	8,28	37,2	8,4	0,81%	1,43%
800	37	6,624	36,8	6,7	0,54%	1,13%
600	36,6	4,968	36,3	5	0,83%	0,64%
400	35,25	3,312	35	3,3	0,71%	0,36%
200	34,1	1,656	33,9	1,65	0,59%	0,36%

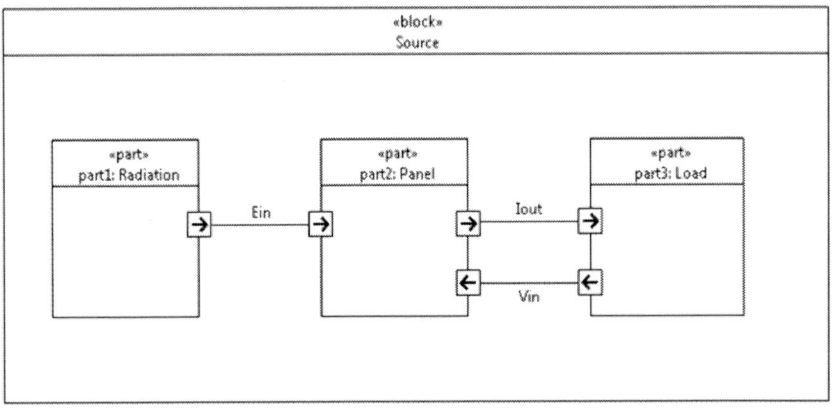

Figure 24. IBD of PV model.

4.3. Maximum Power Point Tracking Technique

Currently there are many different MPPT techniques; these are used in photovoltaic (PV) arrays to obtain the maximum power available from the system. Two configuration types could be used: The first one is that only one algorithm based in machine learning methods is used to the entire array; and this is connected to a unique Boost DC/DC [28][29]. The second one is decentralized the MPPT, ergo, to make a local tracking on each module independently. In any case, the controllers can use only one technique, or switching between multiples algorithms according to the system conditions [30].

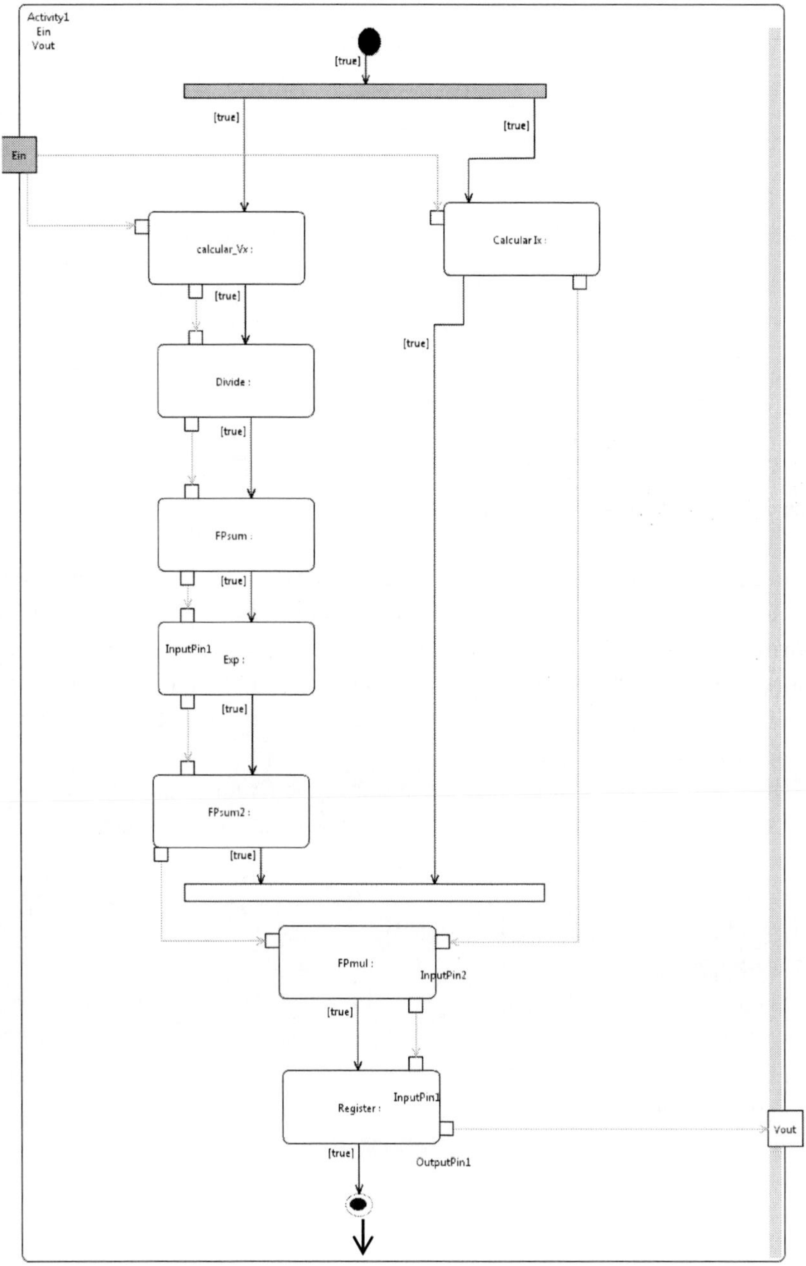

Figure 25. Activity Diagram of PV model.

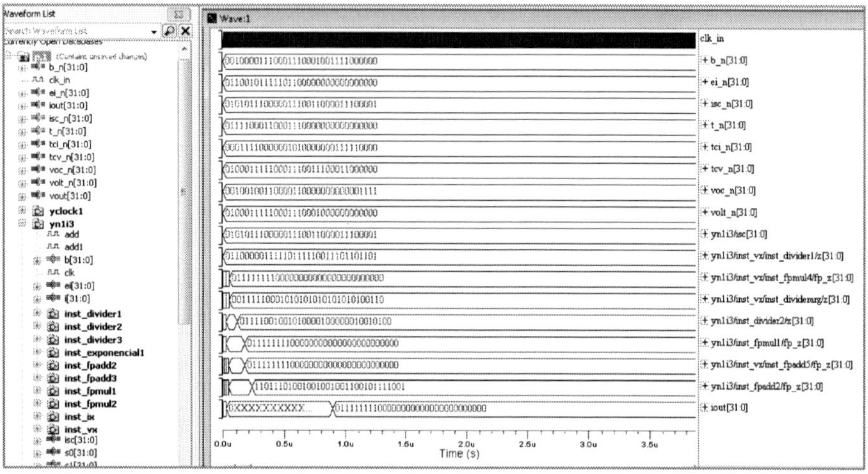

Figure 26. Simulation of PV model in VHDL.

In the most popular algorithms are Perturb and Observe, Incremental Conductance, Current Sweep, Constant Voltage and Slide Control. Many application studies about these methods have been done; and some of them compare several of these techniques using different features like PV array dependent, analog or digital implementation, convergence speed, among others listed in [31][32].

The information collected and the system specifications are used to choose an adequate MPPT technique. In this way, the most optioned algorithms are Perturb and Observe (P&O), Incremental Conductance and Current Sweep. And the main criteria to choice a technique are the difficulty of implementation, the amount and complexity of sensors, the costs and the possible applications.

Based on the above criteria, all techniques selected require sense current and voltage. However, the current sweep is discarded at first, because this depends directly on the PV array and the implementation is more complex than the others techniques. In addition, this method has a slow convergence speed; and the time of current sweep represents loss of available power [32].

In consequence, the P&O is one of the most useful techniques currently. The advantages of this method are: It does not depend on the PV array, it can reach the true MPP, and it has analog or digital implementation with different convergence speeds and less complexity. Nevertheless, this technique has two problems: The first is that despite to reach the MPP, it cannot be maintaining this operating point and generally it oscillates around this; the second is that the method can fail under rapidly changing atmospheric conditions [31][32].

Finally, the Incremental Conductance technique was chosen because it solves the difficulties of the P&O method. The only limitation of this technique is that its implementation is purely digital; however, this is not a major inconvenience nowadays. Using this method it is possible to reach the MPP and maintain this state permanently unless the environmental conditions generate a change in the modules operation, inducing a response from the algorithm. Even under sudden and abrupt changes in irradiation caused by part shadows or other atmospheric changes, the algorithm is able to response in an effectively way setting again the MPP without any risk of fail [31][32].

Nevertheless, the Incremental Conductance technique has two disadvantages. Firstly, it needs a current sensor as the other methods, and this increase considerably the financial costs, even more in this decentralized proposed methodology. Secondly, despite this technique uses different convergence speed, the greater the speed, the more probable is that the MPP cannot be set permanently but an oscillation around it is produced, as explained below.

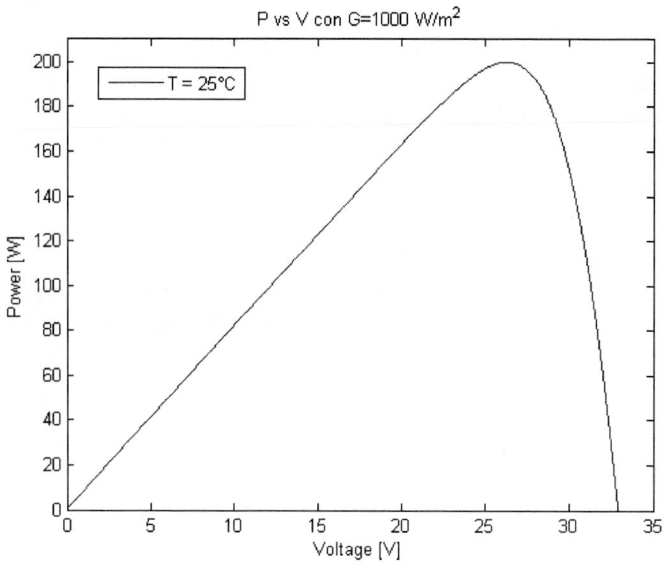

Figure 27. Characteristic curve P vs V of a PV module.

4.3.1. Incremental Conductance

Based on the curve of power vs voltage of a photovoltaic module (figure 9), at the point of maximal power the following equation is satisfied:

$$\frac{dP}{dV} = 0 \tag{29}$$

And since $P = V * I$

$$\frac{d}{dV}(V * I) = 0 \tag{30}$$

$$I + V\frac{dI}{dV} = 0 \tag{31}$$

$$\frac{dI}{dV} = -\frac{I}{V} \tag{32}$$

Then at the MPP, the rate change in output conductance should be equal to the negative output conductance instantaneous. The action of this technique is over the duty cycle (D) of a PWM signal that controls the Boost DC/DC. And to determine the change of this signal, the method is based on the algorithm of the flow chart show in Figure 28.

As can be seen, the algorithm evaluate if the operating point respect to the voltage value is the MPP, some point before it, or some point after it. The algorithm keeps the D value constant when the module operates at the MPP. If the operating point is before the MPP, an increment in the reference voltage is necessary; and this is done by a decrement in the D value. Conversely, if the operating point is after the MPP, a decrement in the reference voltage is necessary; and this is done by an increment in the D value.

The magnitude of the duty cycle step is decided arbitrarily, and that value is responsible of two different behaviors. First, if the increment is higher, the convergence time is fast; but probably the algorithm does not reach exactly the MPP and instead it could oscillate around this point. Second, if the increment is small enough, the algorithm probably reaches the MPP; however, the convergence time will be higher, but this time still remains on the order of milliseconds.

Table 3. Data for the test simulation of the Incremental Conductance algorithm

Interval	V (v)	V0 (v)	I (A)	I0 (A)	dV (v)	dI (A)	-I/V Ω^{-1}	dI/dV Ω^{-1}	D0	D
1	28	0	8	0	28	8	-0,3	0,3	200	0
2	28	28	8	8	0	0	-0,3	inf	0	0
3	28	28	7	8	0	-1	-0,3	inf	0	200
4	28	28	7	7	0	0	-0,3	inf	200	200
5	28	28	7	7	0	0	-0,3	inf	200	200
6	29	28	6	7	1	-1	-0,2	-1,0	200	400
7	29	29	6	6	0	0	-0,2	inf	400	400
8	29	29	5	6	0	-1	-0,2	inf	400	600
9	29	29	5	5	0	0	-0,2	inf	600	600
10	29	29	6	5	0	1	-0,2	inf	600	400
11	28	29	6	6	-1	0	-0,2	0,0	400	400
12	28	28	7	6	0	1	-0,3	inf	400	200
13	28	28	7	7	0	0	-0,3	inf	200	200
14	28	28	8	7	0	1	-0,3	inf	200	0
15	28	28	8	8	0	0	-0,3	inf	0	0

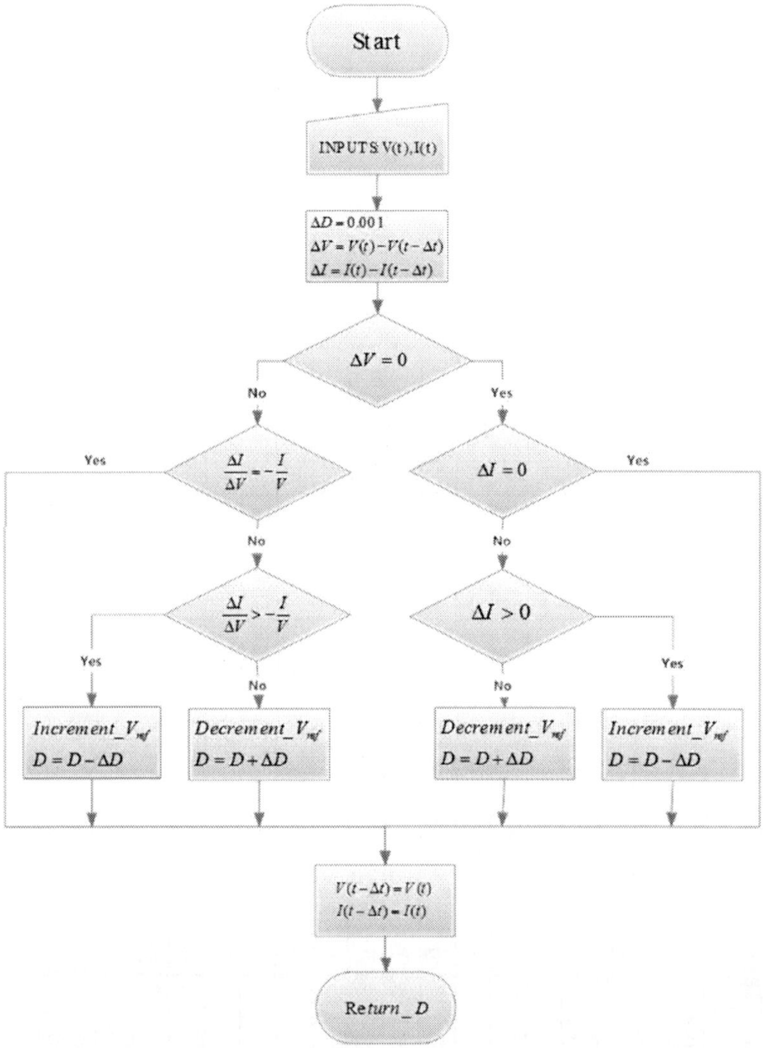

Figure 28. Incremental Conductance Algorithm [32].

4.3.2. MPPT model in VHDL

The Incremental Conductance algorithm is developed in VHDL. To check the functionality, the sequence of voltage (*V*) and current (*I*) values listed in the table III are entered. In that table, the values of other different signals/quantities involved in the process are listed too. These include the actual (D0) and the subsequent (D) duty cycle of the PWM signal; both quantities are given in clock cycles. In the simulation, the clock signal (*clk*) period is 10ns and the sensing signal (*sen*) period is 30μs, this means that every 30μs new values of voltage and current are entered. The PWM signal (*s_pwm*) frequency to control the boost converter is set to 100 KHz, this is equivalent to a period of 10μs or 1000 clock cycles (hence, the D value is given in clock cycles).

Figure 29 shows the corresponding simulation to the data listed in table III. In Figure 29, for each sensing interval there are 3 cycles of the PWM signal. The values of D0 and D are represented by the signals *auxd0* and *auxd*, respectively. The algorithm's response is synchronized to the signal *sen*; and consequently, the delay in the decision process of the change of D makes this change visible only to the subsequent sensing cycle.

To the test simulation (Figure 29) several values of I and V were selected, such that the 3 possible responses could be evidenced. These responses are: the D value is not changed, the D value is increasing, or the D value is decreasing. According to Figure 29, in the interval between left and central markers is possible see the case for a response that keeps the D value constant, in this case at 40%. The beginning of the interval 7 is just where the left marker is positioned (see values in table III); and in that moment, the algorithm's decision is to keep the value of the duty cycle constant, because there is no change in V or I compared to their previous values. Therefore, in the next sensing cycle the D value is 400 clock cycles again, and 3 new pulses of the PWM signal are generated with that value. Similar behavior could be seen in the interval between central and right markers, but in that case the D value is 60%.

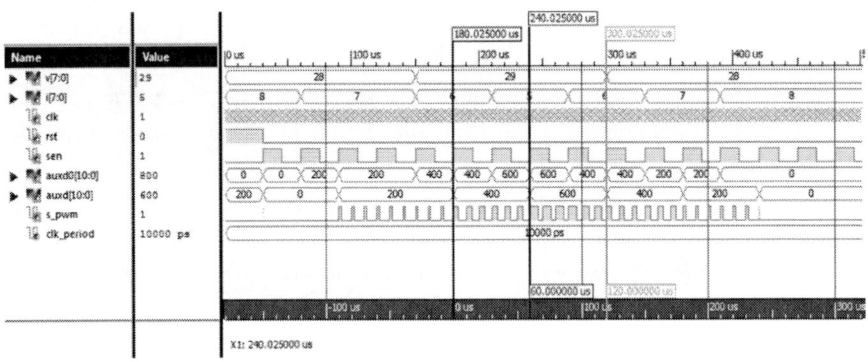

Figure 29. Test simulation of the Incremental Conductance algorithm.

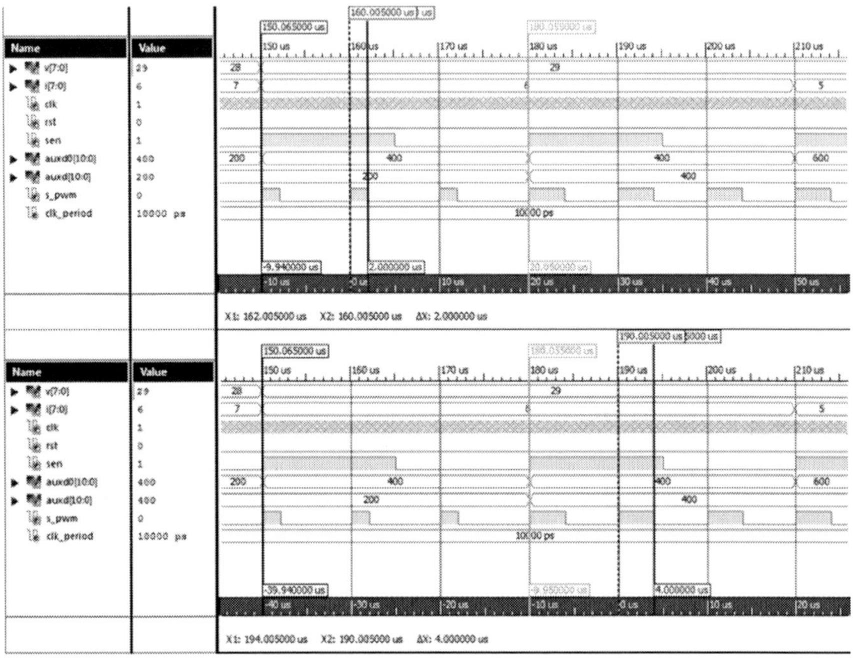

Figure 30. Simulation of the Incremental Conductance algorithm for a response of increasing in the duty cycle of the PWM signal.

In the Figure 30, the algorithm's behavior in case of a response that increases the value of the duty cycle could be noticed. The beginning of the interval 6 is just where the left marker is pointed (see values in table III); and in this moment, a change in V and I compared to their previous values is shown. And in this case, the decision is increasing the value of the duty cycle, because the change in conductance is less than the negative of the instantaneous conductance. Therefore, the actual D value corresponding to 20% is increased to 40% for the next sensing cycle. The amount of increase or decrease in D can be defined in the algorithm; in the present simulation, the amount was set in 200 clock cycles (20%).

In the Figure 31, the algorithm's behavior in case of a response that decreases the value of the duty cycle could be noticed. The beginning of the interval 10 is indicated by the left marker (see values in table III); and in this moment, there is no change in V, but there is a change in I, compared to their previous values. And in this case, the decision is decreasing the value of the duty cycle, because the change in I is positive. Therefore, the actual D value corresponding to 60% is decreased to 40% for the next sensing cycle.

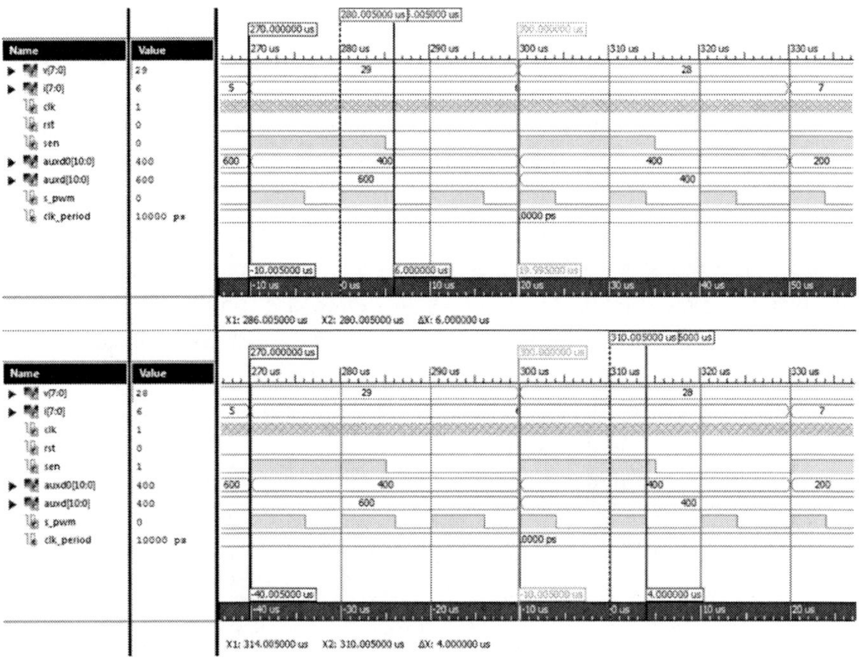

Figure 31. Simulation of the Incremental Conductance algorithm for a response of decreasing in the duty cycle of the PWM signal.

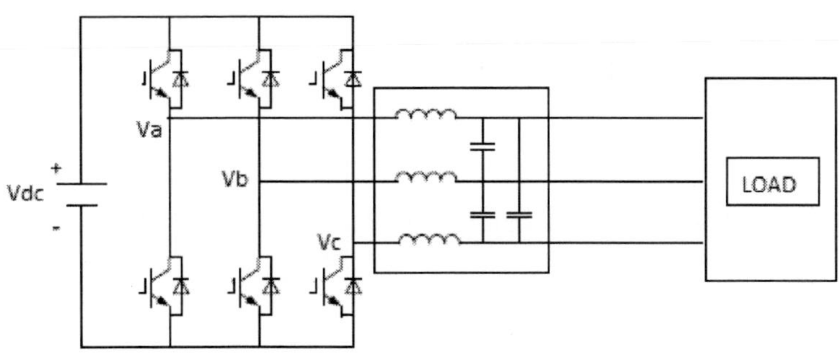

Figure 32. Schematic circuit of three phase inverter.

4.4. Inverter Model

The inverter model implemented in this work was developed according to the model presented in [33]. This model represents the inverter behavior in D-Q and provides the currents and voltages values of the system. Furthermore, this inverter model is intended to regulate the active and reactive power. The proposed model also provides both an active component and a reactive component to the system variables. Finally, the control variables to regulate the system response are the D-Q transformations from the duty cycle of SPWM signals.The schematic circuit of three phase inverter is shown in Figure 32.

The model is given by the equivalent circuit of small signal model, where the inverter is represented by a resistive and a reactive load, whose inputs, the duty cycles of the signals in the upper switches, are modeled by dependent current sources. Which values depends on active and reactive powers and the constant DC input voltage. The values of C,L and w were obtained of the linearization of the model in the steady-state operating point. That model is shown in figure33. And it is deeply described in [33].

The state spaces model of the inverter is given by the following.

$$x = Ax + Bu \tag{33}$$

$$y = Cx \tag{34}$$

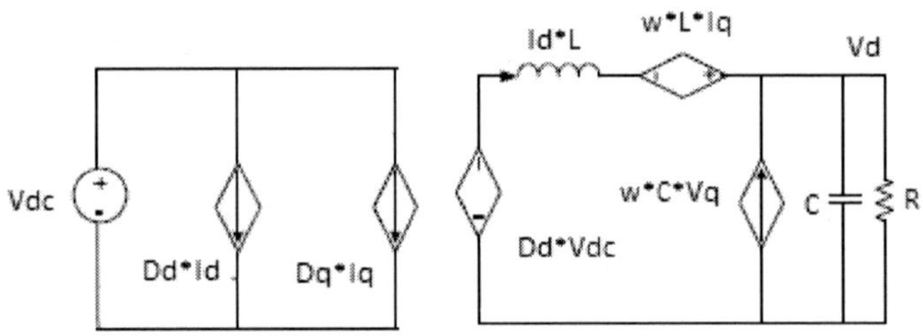

Figure 33. Equivalent circuit of small signal model.

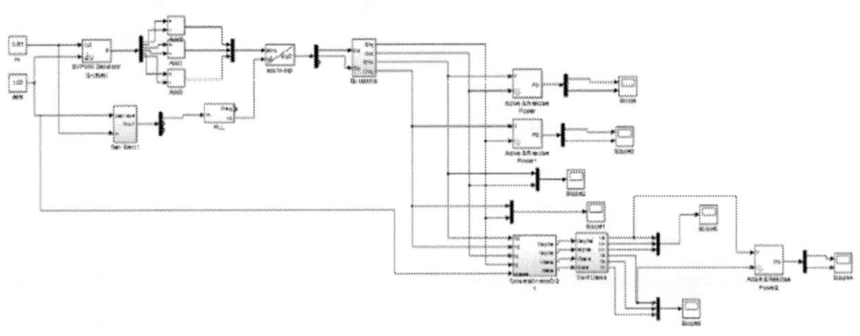

Figure 34. Inverter model.

Where the u vector is composed by the D-Q representation of the duty cycles of the SVPWM control signals (Dd and Dq) and the x vector is composed by the D-Q components of the currents and voltages of the inverter (Id, Iq, Vd, Vq).

$$x^T = [Id\ Iq\ Vd\ Vq] \tag{35}$$

$$u^T = [Dd\ Dq] \tag{36}$$

The values of A,B,C and D coefficients are described by,

$$A = \begin{bmatrix} 0\ w\ \frac{-1}{L}\ 0 \\ -w\ 0\ 0\ \frac{-1}{L} \\ \frac{1}{2C}\ 0\ \frac{-1}{2RC}\ \frac{w}{2} \\ 0\ \frac{1}{2C}\ \frac{-w}{2}\ \frac{-1}{2RC} \end{bmatrix} \tag{37}$$

$$B = \begin{bmatrix} \frac{Vdc}{L}\ 0 \\ 0\ \frac{Vdc}{L} \\ 0\ 0 \\ 0\ 0 \end{bmatrix} \tag{38}$$

$$C = \begin{bmatrix} 1\ 0\ 0\ 0 \\ 0\ 1\ 0\ 0 \\ 0\ 0\ 1\ 0 \\ 0\ 0\ 0\ 1 \end{bmatrix} \tag{39}$$

As the outputs of the system are the currents and voltages in the D-Q space, an inverse Clarke filter was added [34]. It allows obtaining the A, B, C components of the currents and voltages of the system, which would be connected to the other components of the global system. The output filter applied in this case is described by the following conversion matrix.

$$\begin{bmatrix} V_a \\ V_b \\ V_c \end{bmatrix} = \sqrt{\frac{2}{3}} \begin{bmatrix} 1\ 0 \\ -0.5\ \frac{\sqrt{3}}{2} \\ -0.5\ \frac{\sqrt{3}}{2} \end{bmatrix} \begin{bmatrix} V_\alpha \\ V_\beta \end{bmatrix} \tag{40}$$

Where V_α and V_β are given by the following matrix

$$\begin{bmatrix} V_\alpha \\ V_\beta \end{bmatrix} = \begin{bmatrix} cos\phi & -sen\phi \\ sen\phi & cos\phi \end{bmatrix} \begin{bmatrix} V_d \\ V_q \end{bmatrix} \quad (41)$$

And ϕ is the phase angle.

The inverter is modeled to assess its open loop behavior as shown in Figure 34.

The constant values used in the simulation were set as follows.

L=0.660mH
C=0.022mF
R=5Ω
VDC=170V
w=2*pi/1800

The Dd and Dq values were generated as D-Q representation of two signals modulated trough sinusoid pulse width and a phase angle fixed to 100, the maximum pulse width chosen to observe the open loop response is set to 0.91. In Figure 35 are shown the the D-Q representation of the duty cicles of the input signals.

In Figure 36 are shown the values achieved by the D-Q currents, the D-Q voltages, reactive and active power after the simulation of the systems.

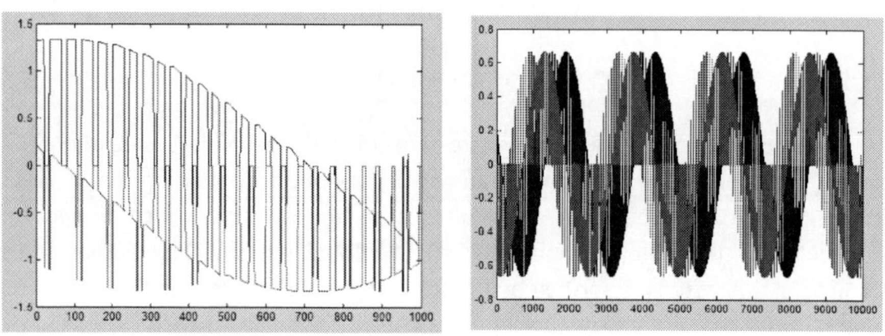

Figure 35. Input signals.

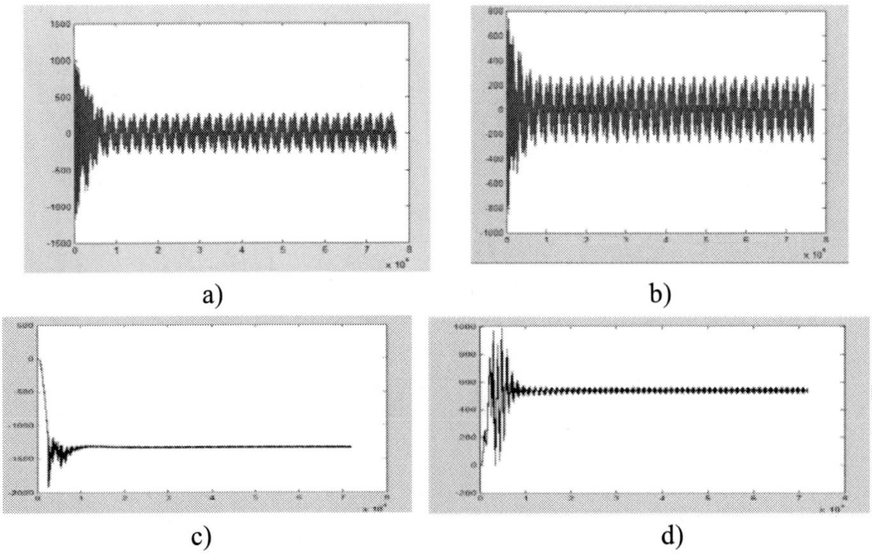

Figure 36. a) Current and Voltage in D coordinate signals, b) Current and voltage in Q coordinate signals, c) Reactive power, d) Active power.

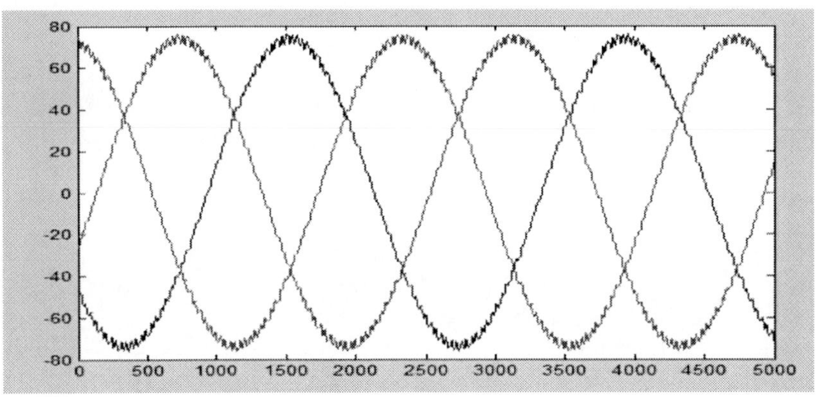

Figure 37. Phase voltages simulated.

The simulations results under open loop system response can be analyzed, observing the values of the phase currents and phase voltage which were obtained when the d-q current values and the d-q voltage values were performed by a filter with the inverse transfer Clarke function. Figure 37 shows the obtained signals.

4.4.1. Inverter design using SYSML and HILES-RCP

The inverter was modeled in HiLeS through the SysML methodology for its implementation in hardware. The description of the system diagram is shown in Figure 38.

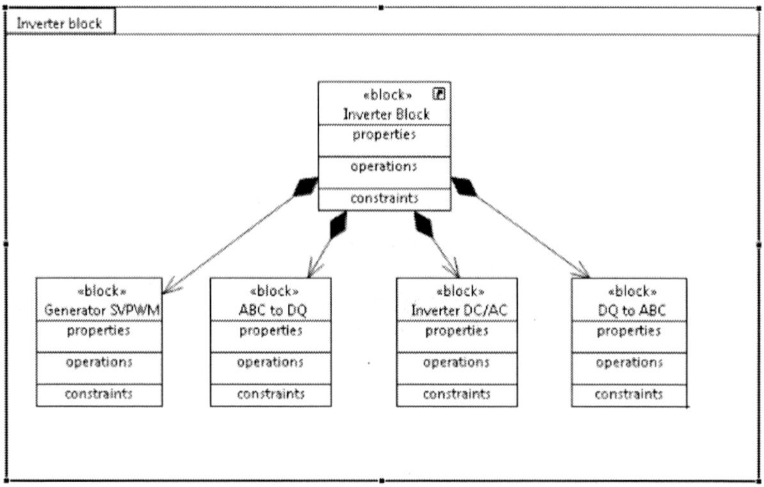

Figure 38. Block definition diagram of inverter model.

The interconnection between the different systems components are defined in the internal block diagram described in the following figure.

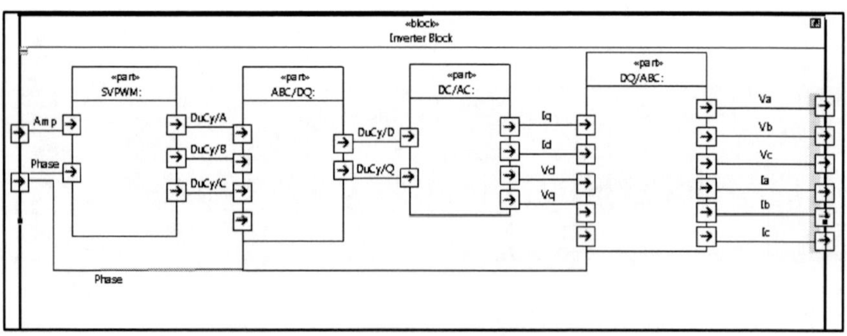

Figure 39. Inverter system model Internal Block Diagram.

In order to model the system behavior is employed an activity diagram whose structure is shown in the following figure.

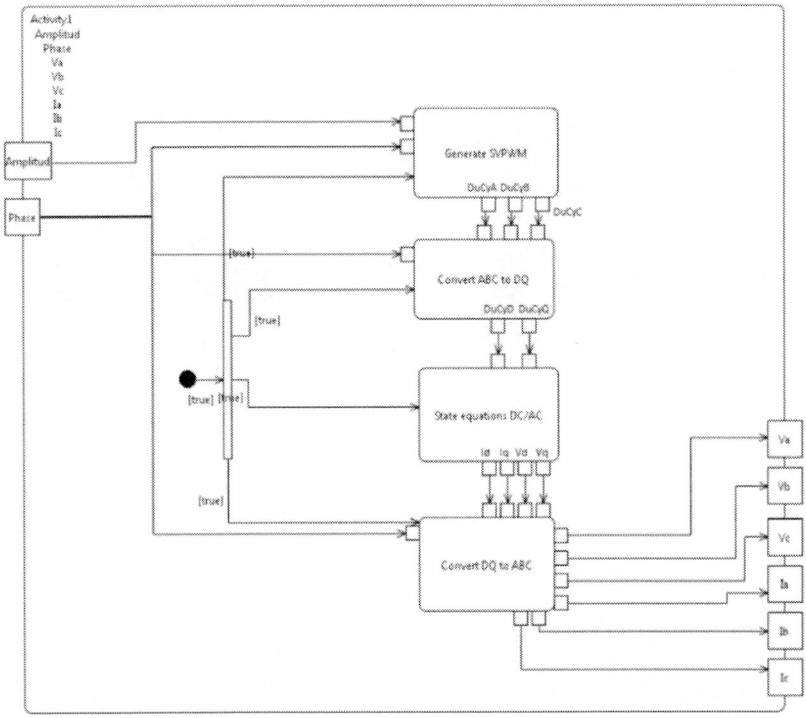

Figure 40. Inverter system model Activity Diagram.

The inverter system in modeled in VHDL and results are shown in the following figures.

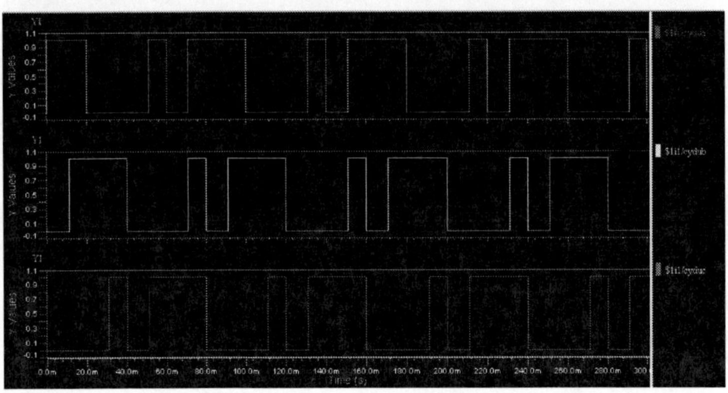

Figure 41. SVPWM signal of inverter modeled in VHDL.

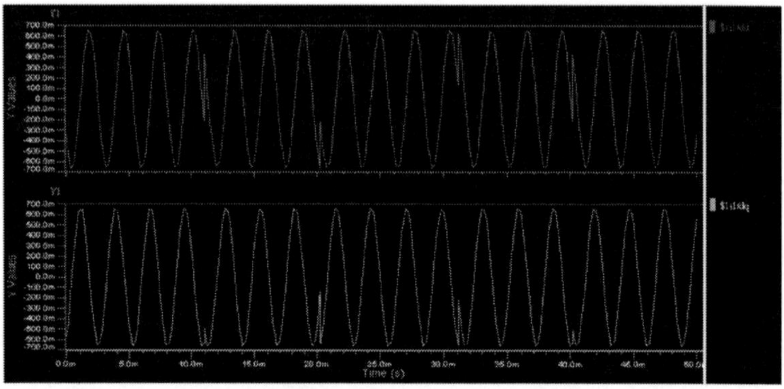

Figure 42. DQ and DD signals of inverter modeled in VHDL.

CONCLUSION

In this chapter the design and implementation of components of microgrids for Hardware-in-the-Loop simulations using SysML and HiLeS-RCP have been presented. The proposed methodology allowed the formal validation of SysML models and their Petri net representations. Also, results confirmed that the model of components of microgrids for Hardware–in-the-Loop simulations showed characteristics equivalents to models using conventional simulation methods. In addition, the methodology proposed to implement the HIL models in FPGA using HiLeS-RCP platform shows successful results. The implementation results showed that HiLeS-RCP is suitable to be used in HIL applications of power electronics for microgrids.

REFERENCES

[1] Bhattacharya, S. K. (2009). *Fundamentals of Power Electronics*, Vikas Publishing House, New Delhi, 11.

[2] Jamil, M., Hussain, B., Abu-Sara, M., Boltryk, R. J. & Sharkh, S. M. (2009). *Microgrid Power Electronic Converters: State of the Art and Future Challenges*, Universities Power Engineering Conference (UPEC), Glasgow.

[3] Blaabjerg, F., Chen, Z., & Kjaer, S. (2004). Power Electronics as Efficient Interface in Dispersed Power Generation Systems, IEEE Trans. *Power Electron.*, vol. *19*, no. 5, 1184–1194.

[4] Midtsund, T. (2010). Control of Power Electronic Converters in Distributed Power Generation Systems, Master Thesis in Energy and Environment, Norwegian University of Science and Technology.

[5] Monmasson, E., Ahoucine, L. & Ohamed, M. (2011). *FPGA-Based Controllers - Different Perspectives of Power Electronics and Drive Applications*, IEEE Industrial Electronics Magazine, 14-26.

[6] Myaing, A. & Dinavahi, V. (2011). FPGA-based real-time emulation of power electronic systems with detailed representation of device characteristics, *IEEE Trans. Ind. Electron.*, vol. *58*, no. 1, 358–368.

[7] Monmasson, E. & Cirstea, M. (2007). FPGA Design Methodology for Industrial Control Systems - A review, *IEEE Trans. Ind. Electron*, vol.*54*, no.4, 1824-1842.

[8] Hoyos, H., Casallas, R., & Jiménez, F. (2012). HiLeS-T: an ADL for Early Requirement Verification of Embedded Systems, in Proceedings of the 5th International Workshop on Model Based Architecting and Construction of Embedded Systems, New York.

[9] HiLeS project. Universidad de los Andes, Colombia, Available in: https://hiles.uniandes.edu.co/

[10] Gomez, C., Esteban, P., Pascal, J. & Jimenez, F. (2009). *Modeling complex system using SysML*, Report LAAS N 09636.

[11] Yoo, C., Choi, W., Chung, I., Won, D., Hong, S. & Jang, B. (2012). *Hardware-in-the-Loop Simulation of DC Microgrid with Multi-Agent System for Emergency Demand Response*, IEEE Power and Energy Society General Meeting, San Diego.

[12] Jeon, J., Kim, J., Kim, H., Kim, S., Cho, C., Kim, J., Ahn, J., & Nam, K. (2010). Development of Hardware In-the-Loop Simulation System for Testing Operation and Control Functions of Microgrid, *IEEE Transactions on Power Electronics*, Vol. *25*, Issue. 12.

[13] Li, Y., Vilathgamuwa, D. M. & Loh, P. C. (2004). Design, Analysis, and Real-Time Testing of a Controller for Multibus Microgrid System, *IEEE Trans. Power Electron.*, vol. *19*, no. 5, 1195-1204.

[14] Hoyos, H., Casallas, R. & Jiménez, F. (2012). *HiLeS-T: an ADL for Early Requirement Verification of Embedded Systems*, in Proceedings of the 5th International Workshop on Model Based Architecting and Construction of Embedded Systems, New York.

[15] Berthomieu, B. & Vernadat, F. (2006). *Time Petri Nets Analysis with TINA*, Third International Conference on Quantitative Evaluation of Systems, 123-124.

[16] David, R. & Alla, H. (2010). *Discrete, Continuous, and Hybrid Petri Nets*, 2nd Edition, Springer, New York.
[17] Jimenez, F. (2000). *Specification et Conception de Micro-Systemes Bases sur des Circuits Asynchrones*, PhD thesis, Institute National des Sciences Appliquees (Toulouse, France), Universidad de los Andes (Bogota, Colombia).
[18] Hoyos, H., Jiménez, F., Casallas, R. & Correal, D. (2013). *Tucan - Virtual Prototype Generation and Time Constraints Analysis of Real-Time Embedded Systems*, Design Automation for Embedded Systems, Springer, New York.
[19] Object Management Group, (2010). *Systems Modeling Language, SysML.*
[20] Bouabana-Tebibel, T., Rubin, S. H. & Bennama, M. (2012). Formal modeling with SysML, IEEE 13th International Conference on Information Reuse and Integration (IRI), 340-347.
[21] Hoyos, H., Casallas, R., Jimenez, F. & Correal, D. (2011). *HiLeS2: Model Driven Embedded System Virtual Prototype Generation*, In Proceedings of the Symposium on Theory of Modeling & Simulation: DEVS Integrative M&S Symposium, San Diego, CA, USA.
[22] Erickson, R. W. (1997). *Fundamentals of Power Electronics*, Chapman and Hall. New York. Second Edition.
[23] Jorge Enrique Salamanca, (2012). *Highly Efficient Photovoltaic Cells and Solar-Panel system at CUBESAT Colombia*, Universidad Distrital de Colombia.
[24] Eduardo, I. & Ortiz, F. Z. (2005). Peng, Analytical Model for a Photovoltaic Module using the Electrical Characteristics provided by the Manufacturer Data Sheet, IEEE, (2005).
[25] 630 SOLAREX Court, *MSX-5 and MSX-10 Photovoltaic Modules*, SOLAREX, (1998).
[26] Siliken Modules, "*SLK60M6L*", SILIKEN S.L.U, (2008).
[27] Abdelsalam, A. K., Massoud, A. M., Ahmed, S. & Enjeti, P. (2011). High-performance adaptive perturb and observe MPPT technique for photovoltaic-based microgrids. *IEEE Transactions on Power Electronics*, *26*(4), 1010-1021,
[28] Shimizu, T., Hirakata, M., Kamezawa, T. & Watanabe, H. (2001). Generation control circuit for photovoltaic modules. *IEEE Transactions on Power Electronics*, *16*(3), 293-300.

[29] Rahmani, R. & Seyedmahmoudian. M. (2013). *Implementation of fuzzy logic maximum power point tracking controller for photovoltaic system.* 10(3), 209-218.

[30] Hohm, D. P. & Ropp, M. E. (2003). *Comparative study of maximum power point tracking algorithms.* Progress in Photovoltaics: Research and Applications, *11*(1), 47_62.

[31] Esram, T. & Chapman, P. L. (2007). *Comparison of photovoltaic array maximum power point tracking techniques.* IEEE Transactions on Energy Conversion, 22(2):439_449.

[32] Zhang, Y., Jiang, Z. & Yu, X. *Small-Signal Modeling and Analysis of Parallel-Connected Voltage Source Inverters*, Department of Electrical and Computer Engineering.University of Miami Coral Gables, USA.

[33] Avendaño, L. E., González, J. P. (2013). *Simulation of a vector control to an induction motor*, IScientia et Technica Año XVIII, Vol. *18*, No 1.

In: Hydroelectricity and Power Electronics
Editor: Lucille Burke
ISBN: 978-1-63321-728-7

Chapter 5

CASCADE CONVERTERS FOR WIDE CONVERSION RATIOS

J. Leyva-Ramos [*] ***and M. G. Ortiz-Lopez***
Instituto Potosino de Investigación Científica y Tecnologica
Universidad Politécnica de San Luis Potosi, Mexico

ABSTRACT

A great number of new technological applications require DC–DC converters with wide range of conversion ratios; for example, processing energy from sustainable DC energy sources. These energy sources, before being connected to the grid, require converters with a significant step-up conversion ratio. On the other hand, there are technological developments that require converters with high-input voltage and low-voltage/high-current output specifications. For example, the new generation of microprocessors is being designed to operate using power supplies with an output voltage from 3.3 V to 1 V or less. A first approach to provide a significant voltage conversion is to operate a switching converter at a very high or small duty cycle. However, the maximum and minimum attainable conversion ratios for the conventional converters are limited by the characteristics of the switching devices. An often-used approach is to use step-down or step-up transformers; however, large switching surges due to leakage inductance may damage the switching devices and make the controller difficult to design. A transformerless topology that provides

[*] Corresponding Author: jleyva@ipicyt.edu.mx

a wide conversion ratio is the cascade converter, which consists in two or more basic switch-mode DC–DC converters connected in cascade where the voltage ratio is a function of the number of stages. In this chapter, a discussion of the operation of cascade converters with a single active switch is given. The operating conditions are developed together with the corresponding models. The resulting models are compatible with the standard techniques widely used in control design.

1.1. INTRODUCTION

Several factors like a growing world concern about global warming, energy security and improvements in technologies [1,2] are requiring DC–DC converters with a wide range conversion voltage. Power processors are required for interfaces to be used with sustainable DC energy sources like photovoltaic modules, batteries of various chemistries and fuel cell. These DC energy sources are all series and parallel connections of basic cells and operate at a low DC voltage ranging from less than 1 V (PV cell, single fuel cell) and 3 V to 24 V (Li-Ion cell). These low voltages do not interface very well to existing higher power systems, so the cells are series connected to create modules with higher terminal voltages. These modules, before being connected to the grid, require converters with significant step-up conversion voltage ratios [3 -9].

An interesting example can be found in the automotive industry that a change is required from the conventional 12 V battery system to a higher voltage, due to increase of electrical load of automotive vehicles that now are estimated in 1 kW for a conventional vehicle to 6 kW in the most luxurious. Manufacturers estimated a growth to 10 kW load in conventional vehicles in the upcoming years.

Other example is in the development of new microprocessors that require innovations in the design of power supplies. The new generation of microprocessors is designed with capacities and speed of operation increasingly great. Currently, requirements of the output voltage produced by a VRM (Voltage regulator module) is in a range of 3.3 V to 1 V, meanwhile the current required is increasing in values ranging from 50 A to 150 A [10]. For the control of the above systems, it is very important to have a mathematical model that describes the dynamic behavior of converters such that different control strategies can be tested in order to design regulators with specific characteristics.

Modelling of converters is required for analysis and controller design. In the power electronics literature, various methods have been used as sampled data modeling [11], average state space modeling [12-14], switching state space modeling [14], averaged PWM switching modeling [15], average generalized modeling [12,13], dynamic phasors modeling [16] and signal flow graph [17 -19]. In this work, switching state space models are developed, resulting in average models.

The reminder of this chapter is organized as follows. In Section 2, conventional converters are studied as a first approach to step-up or step-down a DC voltage; the average state space nonlinear model and average state space linear model are developed.

Three interesting topologies for a wide range conversion are discussed in section 3. In section 4, the dynamical behavior of quadratic buck and boost converters are represented in their state space average nonlinear and linear models. In section 5, a detailed description of operation of cascade buck and boost converters is carried out as well as the expressions to obtain the corresponding ripples for the inductor currents and capacitor voltages. For cascade converters a detailed procedure to obtain models as well as switching state space linear models, average state space nonlinear models and linear models are developed in section 6, 7 and 8 respectively. In section 9, a summary of the models is given. Final remarks are given in section 10.

1.2. CONVENTIONAL CONVERTERS

The operation of a switching converter is based on a fast switching of active switch from the ON condition to the OFF condition and vice versa, that allows charging a capacitor and an inductor, which results an output voltage as a function of a nominal duty ratio U. This is the average value of the fraction of time that active switch is ON of a control signal with period T. A topology that provides a down conversion voltage is the buck converter and for the up conversion voltage is the boost converter as shown in Figure 1.1. Thus in these circuits as those shown hereinafter, active switch and diodes are ideal switches, then, there are not power loses and they only open or closed.

In these converters, the input voltage is represented by E and output voltage by V_0. The inductor and capacitor of the converter are given by L and C, respectively and the load modeled by R. The energy is transferred to the output using a MOSFET as a controlled switch represented by S and by a diode represented as D.

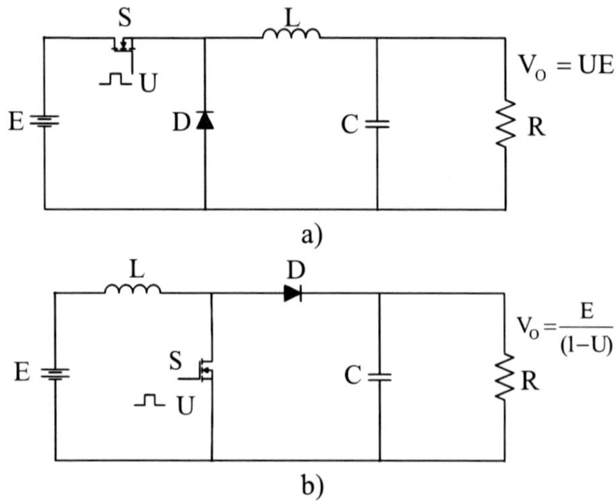

Figure 1.1. Conventional converters: a) buck converter; and b) boost converter.

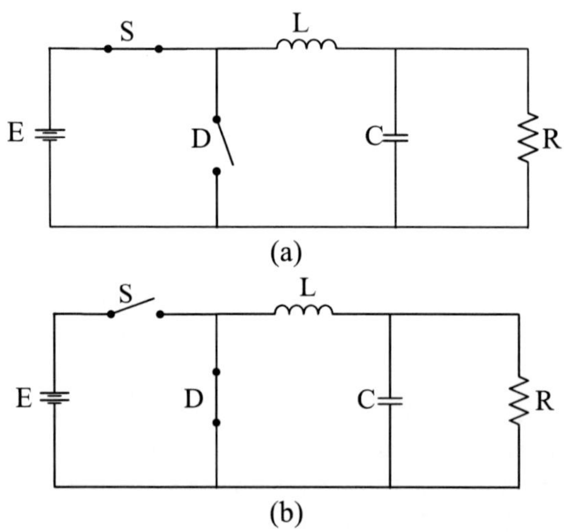

Figure 1.2. Conventional buck converter: a) active switch ON, and b) active switch OFF.

The models of the converters using the state-space equations are derived from each circuit that results when active and passive switches are ON or OFF using Kirchhoff equations. For the buck converter, the two resulting circuits are shown in Figure 1.2.

The electrical circuit can be obtained when active switch is ON, and therefore passive switch is OFF $(0 \le t \le UT)$, where the states are inductor current $i_L \,\square\, x_1$ and capacitor voltage $v_C \,\square\, x_2$ respectively; thus, state-space equations are given by:

$$\dot{x} = \begin{bmatrix} 0 & -\frac{1}{L} \\ \frac{1}{C} & -\frac{1}{RC} \end{bmatrix} \begin{bmatrix} x_1 \\ x_2 \end{bmatrix} + \begin{bmatrix} \frac{1}{L} \\ 0 \end{bmatrix} E \tag{1.1}$$

When active switch is OFF, and therefore passive switch is ON $(UT \le t \le T)$, the state-space equations are:

$$\dot{x} = \begin{bmatrix} 0 & -\frac{1}{L} \\ \frac{1}{C} & -\frac{1}{RC} \end{bmatrix} \begin{bmatrix} x_1 \\ x_2 \end{bmatrix} + \begin{bmatrix} 0 \\ 0 \end{bmatrix} E \tag{1.2}$$

For the buck converter, a switching function q can be defined for the period of time when active switch is ON. From the state-space equations and by using the switching function defined in the former, the switching state-space model in a matrix representation is given by:

$$\dot{x} = \begin{bmatrix} 0 & -\frac{1}{L} \\ \frac{1}{C} & -\frac{1}{RC} \end{bmatrix} \begin{bmatrix} x_1 \\ x_2 \end{bmatrix} + \begin{bmatrix} \frac{q}{L} \\ 0 \end{bmatrix} E \tag{1.3}$$

For modeling the dynamics of boost converter, the resulting electrical circuits when active switch is ON and OFF are analyzed. These electrical circuits are shown in Figure 1.3.

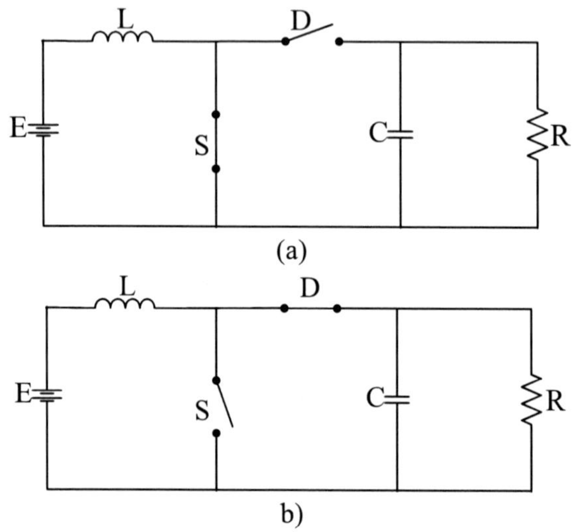

Figure 1.3. Conventional boost converter: a) active switch ON, and b) active switch OFF.

The resulting state-space equations when the active switch is ON, and as consequence passive switch is OFF $(0 \le t \le UT)$, are:

$$\dot{x} = \begin{bmatrix} 0 & 0 \\ 0 & -\frac{1}{RC} \end{bmatrix} \begin{bmatrix} x_1 \\ x_2 \end{bmatrix} + \begin{bmatrix} \frac{1}{L} \\ 0 \end{bmatrix} E \tag{1.4}$$

For active switch OFF, and as consequence passive switch ON $(UT \le t \le T)$, the state-space equations are:

$$\dot{x} = \begin{bmatrix} 0 & -\frac{1}{L} \\ \frac{1}{C} & -\frac{1}{RC} \end{bmatrix} \begin{bmatrix} x_1 \\ x_2 \end{bmatrix} + \begin{bmatrix} \frac{1}{L} \\ 0 \end{bmatrix} E \tag{1.5}$$

Based on the operation of boost converter, a switching function q can be defined, which corresponds to active switch ON. By the use of state space equations derived in the former and using the switching function, the switching state space model of boost converter is given by:

$$\dot{x} = \begin{bmatrix} 0 & -\frac{(1-q)}{L} \\ \frac{(1-q)}{C} & -\frac{1}{RC} \end{bmatrix} \begin{bmatrix} x_1 \\ x_2 \end{bmatrix} + \begin{bmatrix} \frac{1}{L} \\ 0 \end{bmatrix} E \tag{1.6}$$

Average modeling technique is one of the most important tools for converter analysis used in power electronics. Average behavior of a topology provides important information about converter performance in both DC conditions as in low frequency, regardless of ripple, switching and other fast effect. Even when ripple is not taken into consideration in the average output, this representation is very useful to determine transient response and steady state regulation. Also average modelling allows large changes in voltage and current of source and load, changes of control inputs and robust analysis in the presence of parameters changes in circuit components. Additionally, the ripple effect can be considered at the end of modeling process. The main objective of average technique is to find an approximate circuit to analyze local average behavior of circuit variables, even during transient response, which is a condition with no periodical characteristics [15].

Average local value of a variable is defined as:

$$\overline{x} = \frac{1}{T_S} \int_{t_i}^{t_i+T_S} x(\tau)\, d\tau \tag{1.7}$$

where T_S is a fixed value, which represents the period of the function $x(\tau)$ and t_i represents the time where averaging begins. It is very important to choose a suitable value of T_S, according to the case in which is applied, to obtain correct results.

For switching converters, value of T_S is equal to the inverse of switching frequency of converter, and t_i coincides with the beginning of switching period. Function $q(t)$ takes binary value of one, in the period of time between t_i and $t_i + uT_S$, and a binary value of zero, in the period of time between $t_i + uT_S$ y $t_i + T_S$, as is shown in Figure 1.4.

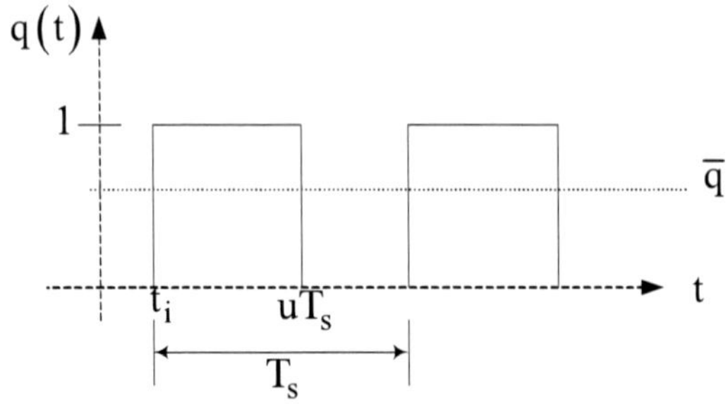

Figure 1.4. Average of switching function q(t).

Average value of switching function, therefore, is given by:

$$\overline{q} = \frac{1}{T_S}\left[\int_{t_i}^{t_i+uT_S}(1)dt + \int_{t_i+uT_S}^{t_i+T_S}(0)dt\right] = \overline{u} \tag{1.8}$$

Average models of the converters can be obtained from the state equations for each configuration produced by commutation switching of active and passive switches, by using of duty ratio instead of switching function. These models provide average circuits that produced the same inputs and outputs as the original model because the sequence of circuits of a converter adds or decreases energy to produce well defined average results at input or output.

The resulting system has a representation of the following form:

$$\dot{\overline{x}} = F(\overline{u})\overline{x} + G(\overline{u})\overline{e} \tag{1.9}$$

where $\overline{x} \in R$ is the state vector and $\overline{e} \in R$ is the input voltage. The matrix F is of dimensions $n \times n$, the column vector G is of dimension n, where n is the number of state variables.

To study of the dynamical behavior of these converters, the concept of average modelling is used and the corresponding models given in the state space matrix representation of the form $\dot{\overline{x}} = F(\overline{u})\overline{x} + G(\overline{u})\overline{e}$, where $\overline{x}$ is the state vector given by $[\overline{i}_L, \overline{v}_C]^T \in \mathbb{R}^2$, F is a $\mathbb{R}^{2x2}$ matrix, G is a $\mathbb{R}^2$ vector, $\overline{e}$ represents input voltage and $\overline{u}$ control signal for active switch.

For a buck converter, the average model is given by:

$$\begin{bmatrix} \dot{\bar{i}}_L \\ \dot{\bar{v}}_C \end{bmatrix} = \begin{bmatrix} 0 & -\frac{1}{L} \\ \frac{1}{C} & -\frac{1}{RC} \end{bmatrix} \begin{bmatrix} \bar{i}_L \\ \bar{v}_C \end{bmatrix} + \begin{bmatrix} \frac{\bar{u}}{L} \\ 0 \end{bmatrix} \bar{e} \tag{1.10}$$

and for a boost converter by:

$$\begin{bmatrix} \dot{\bar{i}}_L \\ \dot{\bar{v}}_C \end{bmatrix} = \begin{bmatrix} 0 & -\frac{(1-\bar{u})}{L} \\ -\frac{(1-\bar{u})}{C} & -\frac{1}{RC} \end{bmatrix} \begin{bmatrix} \bar{i}_L \\ \bar{v}_C \end{bmatrix} + \begin{bmatrix} \frac{1}{L} \\ 0 \end{bmatrix} \bar{e} \tag{1.11}$$

The matrix F and vector G depend on control signal $\bar{u} \in \mathbb{R}$. These are nonlinear representations and only valid for high frequency operated converters in continuous conduction mode (CCM) [15]. In this condition, the inductor current never decays to zero. All the models given in this chapter are for CCM.

In the case of switching converters, the linearized model can be developed as known as small signal model. This model represents the system within a limited operation range but can be used to study the dynamical properties of converters. These linear-invariant models describe approximately the behavior of converters for frequencies up to half of switching frequency; therefore, they are no suitable for predict sub-harmonic oscillations because of ripple instability of voltages and currents.

The following procedure is used to carry out the model linearization:

1. The control parameters and state variables are decomposed in two parts. The first part contains the nominal values. The second one the deviations from the nominal.
2. Equations are rewritten using the above decomposition. Assuming that the deviations are sufficiently small, nonlinear and second order terms are neglected.
3. In the resulting expressions, values of constant parameters are replaced by the nominal value of variables.

These average models can be linearized around an operating point, given expressions of the form $\dot{\tilde{x}} = F\tilde{x} + G\tilde{u}$ as shown in (1.12a) for a buck converter and (1.12b) for a boost converter.

$$\begin{bmatrix} \dot{\tilde{i}}_L \\ \dot{\tilde{v}}_C \end{bmatrix} = \begin{bmatrix} 0 & -\dfrac{1}{L} \\ \dfrac{1}{C} & -\dfrac{1}{RC} \end{bmatrix} \begin{bmatrix} \tilde{i}_L \\ \tilde{v}_C \end{bmatrix} + \begin{bmatrix} \dfrac{E}{L} \\ 0 \end{bmatrix} \tilde{u} \tag{1.12a}$$

$$\begin{bmatrix} \dot{\tilde{i}}_L \\ \dot{\tilde{v}}_C \end{bmatrix} = \begin{bmatrix} 0 & -\dfrac{(1-U)}{L} \\ \dfrac{(1-U)}{C} & -\dfrac{1}{RC} \end{bmatrix} \begin{bmatrix} \tilde{i}_L \\ \tilde{v}_C \end{bmatrix} + \begin{bmatrix} \dfrac{E}{(1-U)L} \\ \dfrac{E}{(1-U)^2 RC} \end{bmatrix} \tilde{u} \tag{1.12b}$$

However, the maximum and the minimum attainable conversion ratios for the conventional converters are limited by the characteristics of the switching devices. The turn-ON time and turn-OFF time of the active switch now play an important role for the attainable duty ratio and, consequently, in the conversion ratio. In addition, when duty ratio is very near to 0 or to 1, there are a large signal deterioration of output voltage and inductor current and consequently of the control signal.

1.3. Other Proposed Converter Topologies

There are a wide variety of topologies available in the open literature for higher conversion ratios; however, the most representative are discussed in this section.

Isolated Converters

An often-used approach to step-down or step-up output voltage is to use transformers; however, large switching surges appear because of leakage inductance effects may damage the switching devices and make the controller difficult to design. To avoid this problem, cascade two stage topologies using a conventional converter in first stage and a transformer in the output stage has been proposed as shown in Figure 1.5 [20, 21]. The work in this area has been focus in low voltage-high current converters with applications in new generation of microprocessors and data communications systems.

In Figure 1.5, the active switch of conventional stage S_1 operates with a duty ratio U_1. The active switches of isolated stage are represented as S_2 to S_3. In the first three topologies shown, switches S_2 y S_3 operates with the same duty ratio, but in alternative switching periods.

In the fourth topology, switches S_2 and S_5 operate simultaneously as S_3 with S_4. The four switches operate with the same duty ratio U_2 but each pair working in alternative switching periods. About the second stage schemes, the most used is push-pull topology because it presents low loses in primary stage of transformer, that at all time there is only a MOSFET connected in series with input voltage. As drawback of this topology, there is the problem of saturation of transformer core because it is not be guaranteed that conduction time of MOSFETs S_2 and S_3 is exactly equal; thus, current imbalance could be appear, causing DC component apply to transformer when is not equal to zero. For this reason, each two switching periods could result in an increment of magnetizing current that produces a saturation of transformer core. In addition, this topology presents sub optimization of transformer because primary and secondary winding have a central tap.

Forward topology is appropriate to handle a high current because in the output current there is not oscillations. There are few loses in magnetic core, as drawback it only could be used with a duty ratio that varies between $0 < U \leq 0.5$. Half-bridge topology is recommended to use for power up to 750 W because of their four transistors and their respective operating circuits. In half-bridge topology, the resulting voltage applied to the transformer is a half of that achieve with full-bridge topology; thus, turns number of transformer winding should be double. As consequence, the value of applied current to transistors is twice and; therefore, it is adequate for low power applications. It is noteworthy to notice that because of its physical characteristics, current-mode control could not be implemented in half-bridge topology.

Small-signal models have been carried out for the four topologies with isolated stage and control schemes, where three loops have proposed using in the inner loop an average current-mode control scheme [20].

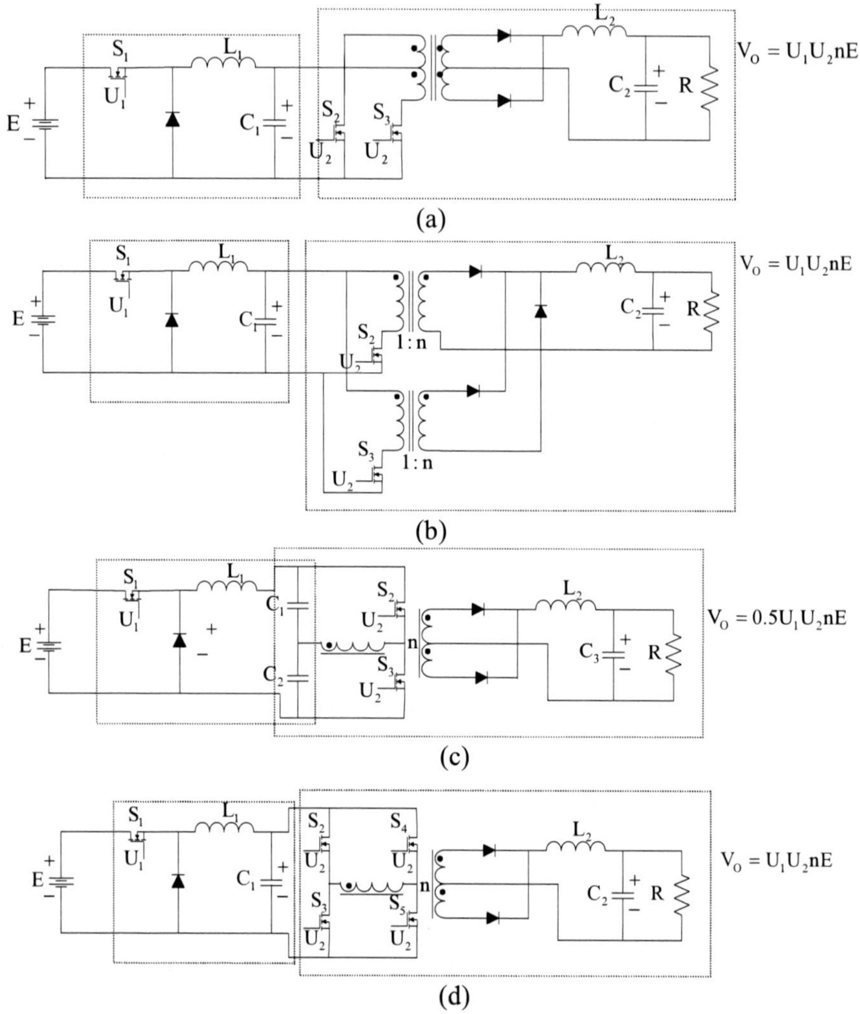

Figure 1.5. Wide ratio output voltage converters with isolated stage: a) buck + push-pull, b) buck + forward, c) buck + half-bridge, and d) buck + full-bridge.

Interleaved Converters

This topology consists of a number of single converters with parallel interconnection between them. The signals to manage them are also interconnected and phase shifted so that there is always a converter working at once, for this reason, this class of converts are also known as multiphase converters. A typical circuit is shown in Figure 1.6 [22 -24].

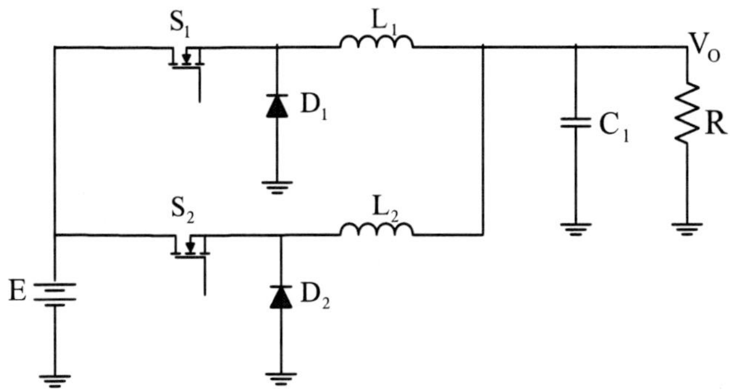

Figure 1.6. Interleaved buck converter.

However, to obtain a wide step-up or step-down voltage ratio, they have to be operated with an extremely down or up duty ratio with the corresponding problems in transient response and efficiency. There are topologies that have been proposed that use better duty ratios, as multi-stage inductor or coupling inductors. A drawback of these topologies is the presence of severe peaks in switching devices because of inductor linkage between windings.

Cascaded Converters

A topology that also provides a wide conversion ratio is cascaded converters, which consists in two or more basic switch-mode DC–DC converters connected in cascade, where the voltage ratio is given as function of the duty ratios U_i [25]. Buck converters in cascade and boost converters in cascade are shown in Figure 1.7. Input voltage-output voltage ratios as a function of nominal duty ratios U_i are given by: a) n-stage buck converters $\prod_{i=1}^{n} U_i$, and b) n-stage boost converters by $\prod_{i=1}^{n} 1/(1-U_i)$.

The dynamic behavior of these converters can be studied using the concept average models. These models can be represented in state space of the following form $\dot{\overline{x}} = F(\overline{u})\overline{x} + G(\overline{u})\overline{e}$, which are valid for switching converters operating at high frequency under continuous conduction mode [15]. These average models can also be linearized around an operating point. These linear time invariant models describe the behavior of the converter for frequencies up to half the switching frequency. Other techniques like diagram flow signal can also be used to model switching converters; thus, the corresponding transfer functions can be easily obtained.

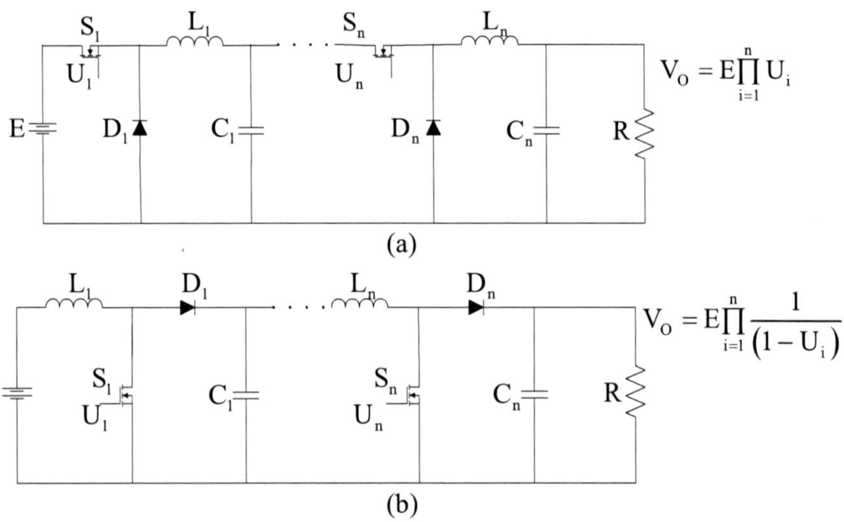

Figure 1.7. N-stage converters: a) buck, and b) boost.

1.4. Quadratic Converters

When a quadratic conversion is required, two stage converters with a single active switch are proposed. In the case of the quadratic buck converter, the modified scheme [27] is shown in Figure 1.8. This converter could be seen as a converter formed by a buck passive stage (L_1, C_1, D_1, D_2) and a buck active stage (L_2, C_2, D_3, S). Switching of diodes inside buck passive stage is a consequence of pulsating current from buck active stage.

Quadratic boost converter with a single switch has been proposed [28 - 30], which is shown in Figure 1.9. The resulting topology could be considerate as the cascade connection of two boost converters. The first one built by the network (L_1, C_1, D_1) and the second one by (L_2, C_2, D_3) where diode D_2 and switch S form the active switch of first stage, and S switch is the active switch of second stage.

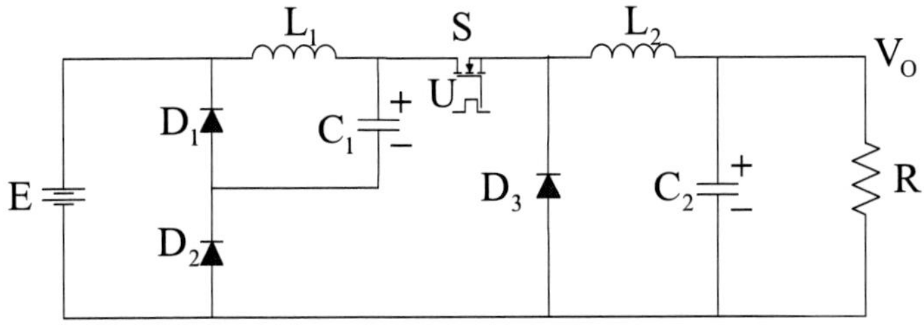

Figure 1.8. Quadratic buck converter.

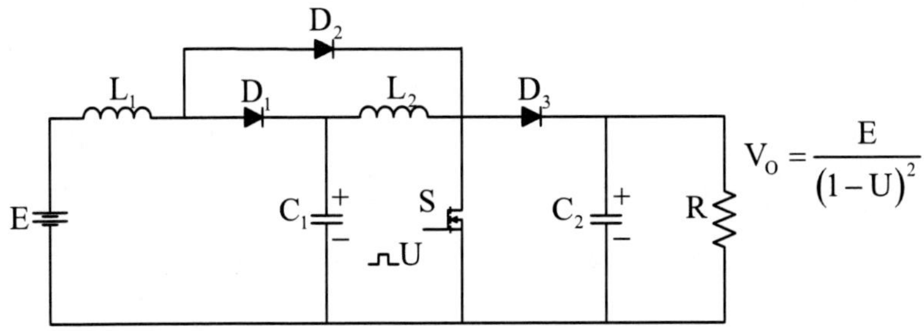

Figure 1.9. Quadratic boost converter.

To study the dynamical behavior of both classes of converts, the concept of average model can be used. The state space representation of the form $\dot{\overline{x}} = F(\overline{u})\overline{x} + G(\overline{u})\overline{e}$ can be obtained where $\overline{x}$ the state vector is given by $[\overline{i}_{L_1}, \overline{i}_{L_2}, \overline{v}_{C_1}, \overline{v}_{C_2}]^T \in \mathbb{R}^4$, matrix F is $\mathbb{R}^{4x4}$, vector G is $\mathbb{R}^4$, and the vector $\overline{e}$ represents input voltage and control signal $\overline{u}$ for active switch. It can be noticed that the state vector contains two inductor currents and two capacitor voltages. Using averaging techniques, the resulting average model [31] for the quadratic buck converter is given by:

$$\begin{bmatrix} \dot{\overline{i}}_{L_1} \\ \dot{\overline{i}}_{L_2} \\ \dot{\overline{v}}_{C_1} \\ \dot{\overline{v}}_{C_2} \end{bmatrix} = \begin{bmatrix} 0 & 0 & -\frac{1}{L_1} & 0 \\ 0 & 0 & \frac{\overline{u}}{L_2} & -\frac{1}{L_2} \\ \frac{1}{C_1} & -\frac{\overline{u}}{C_1} & 0 & 0 \\ 0 & \frac{1}{C_2} & 0 & -\frac{1}{RC_2} \end{bmatrix} \begin{bmatrix} \overline{i}_{L_1} \\ \overline{i}_{L_2} \\ \overline{v}_{C_1} \\ \overline{v}_{C_2} \end{bmatrix} + \begin{bmatrix} \frac{\overline{u}}{L_1} \\ 0 \\ 0 \\ 0 \end{bmatrix} \overline{e} \qquad (1.13a)$$

and for the quadratic boost converter [32] by:

$$\begin{bmatrix} \dot{\bar{i}}_{L1} \\ \dot{\bar{i}}_{L2} \\ \dot{\bar{v}}_{C1} \\ \dot{\bar{v}}_{C2} \end{bmatrix} = \begin{bmatrix} 0 & 0 & \frac{(1-\bar{u})}{L_1} & 0 \\ 0 & 0 & \frac{1}{L_2} & \frac{(1-\bar{u})}{L_2} \\ \frac{(1-\bar{u})}{C_1} & \frac{1}{C_1} & 0 & 0 \\ 0 & \frac{(1-\bar{u})}{C_2} & 0 & \frac{1}{RC_2} \end{bmatrix} \begin{bmatrix} \bar{i}_{L1} \\ \bar{i}_{L2} \\ \bar{v}_{C1} \\ \bar{v}_{C2} \end{bmatrix} + \begin{bmatrix} \frac{1}{L_1} \\ 0 \\ 0 \\ 0 \end{bmatrix} \bar{e} \tag{1.13b}$$

Because matrix F and vector G depend on control signal $\bar{u} \in \mathbb{R}$, these are nonlineal representations.Also, these average models could be linearized around an operating point, which result in expressions in the form $\dot{\tilde{x}} = F\tilde{x} + G\tilde{u}$. In these expressions, the nominal values are denoted by uppercase letters and the deviation from nominal point are denoted by the superscript ~.

For the quadratic buck converter the linearized model is given by:

$$\begin{bmatrix} \dot{\tilde{i}}_{L1} \\ \dot{\tilde{i}}_{L2} \\ \dot{\tilde{v}}_{C1} \\ \dot{\tilde{v}}_{C2} \end{bmatrix} = \begin{bmatrix} 0 & 0 & -\frac{1}{L_1} & 0 \\ 0 & 0 & \frac{U}{L_2} & -\frac{1}{L_2} \\ \frac{1}{C_1} & -\frac{U}{C_1} & 0 & 0 \\ 0 & \frac{1}{C_2} & 0 & -\frac{1}{RC_2} \end{bmatrix} \begin{bmatrix} \tilde{i}_{L1} \\ \tilde{i}_{L2} \\ \tilde{v}_{C1} \\ \tilde{v}_{C2} \end{bmatrix} + \begin{bmatrix} \frac{U}{L_1} \\ \frac{UE}{L_1} \\ -\frac{U^2E}{RC_1} \\ 0 \end{bmatrix} \tilde{u} \tag{1.14a}$$

and for the quadratic boost converter by:

$$\begin{bmatrix} \dot{\tilde{i}}_{L_1} \\ \dot{\tilde{i}}_{L_2} \\ \dot{\tilde{v}}_{C_1} \\ \dot{\tilde{v}}_{C_2} \end{bmatrix} = \begin{bmatrix} 0 & 0 & -\frac{(1-U)}{L_1} & 0 \\ 0 & 0 & \frac{1}{L_2} & -\frac{(1-U)}{L_2} \\ \frac{(1-U)}{C_1} & -\frac{1}{C_1} & 0 & 0 \\ 0 & \frac{(1-U)}{C_2} & 0 & -\frac{1}{RC_2} \end{bmatrix} \begin{bmatrix} \tilde{i}_{L_1} \\ \tilde{i}_{L_2} \\ \tilde{v}_{C_1} \\ \tilde{v}_{C_2} \end{bmatrix} + \begin{bmatrix} \frac{E}{(1-U)L_1} \\ \frac{E}{(1-U)^2L_2} \\ -\frac{E}{(1-U)^4RC_1} \\ -\frac{E}{(1-U)^3RC_2} \end{bmatrix} \tilde{u} \tag{1.14b}$$

1.5. Cascade Buck and Boost Converters

When quadratic converters cannot satisfy requirements of a wider conversion ratio, the above topologies can be extended to n-stages with a single active switch. These topologies are shown in this section with the corresponding average models that can be used later to develop the appropriate controllers.

Cascade Buck Converter

As mentioned previously, in a quadratic buck converter described in section 1.4, the pulsating input current produced by buck active stage is used to switch the diodes of the passive stage. Thus, if $(n-1)$ passive buck stages are connected in cascade in a row with a single switch, a buck cascade converter with a U^n conversion ratio is obtained. This topology is shown in Figure 1.10. As it can be notice, n inductors L_i, n capacitors C_i, a single active switch S and $(2n-1)$ diodes D_i are required. Nominal duty ratio is represent by U, which denotes the percentage of time that active element is ON in a whole switching cycle, thus its value is given by $0 < U < 1$. The input voltage is represented by E, output voltage as V_O, switching frequency as f, load modeled by R and n denote the number of stages of converter.

The conversion ratio is obtained by assuming that converter is operating in CCM. In this operating condition, all capacitor voltages and inductor currents have a DC value with a relatively small value of AC. Besides turn ON and turn OFF times of diodes are synchronized with turn ON and turn OFF time of active switch.

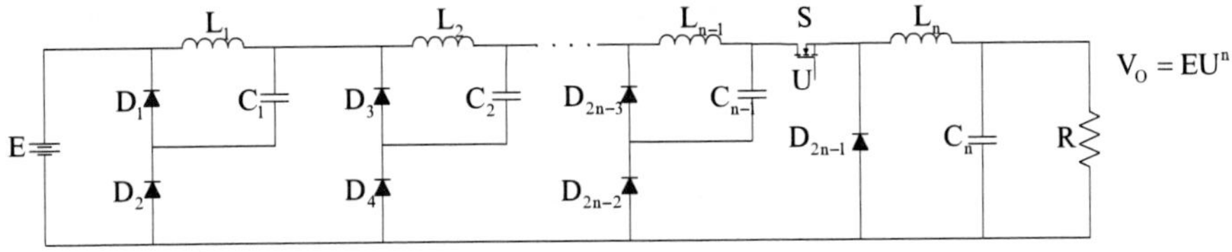

Figure 1.10. Cascade buck converter.

In the buck converter shown in Figure 1.10, when the active switch S is ON, simultaneously diodes D_2, D_4 to D_{2n-2} are ON, thus there is a closed trajectory for currents. At this time D_1, D_3 to D_{2n-1} are OFF and do not allow current flow. The circuit diagram is showed in Figure 1.11.

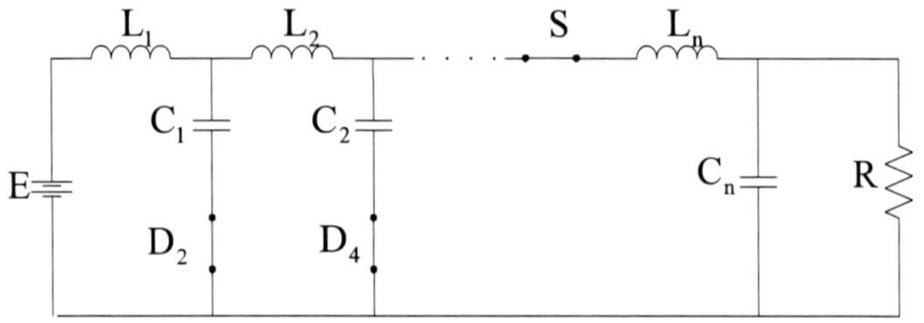

Figure 1.11. Cascade buck convert with the active switch closed.

Later, when the active switch S is closed, during the part of duty cycle corresponding to $(1-U)$, diodes D_2, D_4 to D_{2n-2} turned OFF, and diodes D_1, D_3 to D_{2n-1} turned ON, allowing a new current trajectory as shown in Figure 1.12.

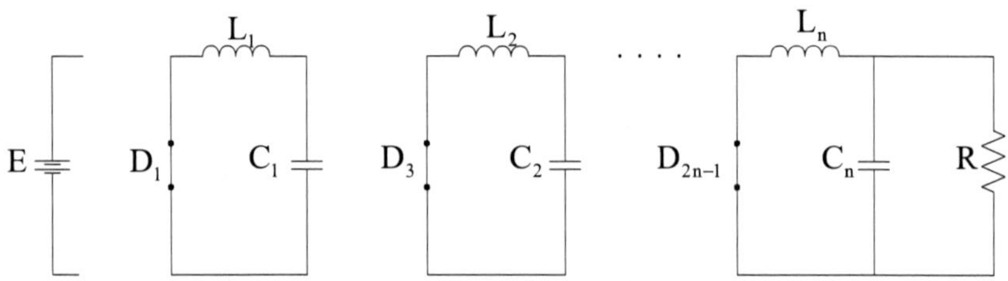

Figure 1.12. Cascade buck converter with the active switch OFF.

The steady state response of former circuits can be obtained using the same techniques applied to conventional converters [14]. It is found that capacitor voltages are given by $V_{Ci} = EU^i$ for $i = 1, \cdots, n$, and inductor currents by $I_{Li} = I_O U^{n-i}$ for $i = 1, \cdots, n$, where I_O is the output current. As can be seen, capacitor voltages decrease and the inductor currents increase throughout the converter. The output voltage is given by $V_O = EU^n$.

Besides, when a converter is designed, it must meet certain specifications as ripple value of capacitor voltages and output voltage, as well as ripple value of inductor currents. These specifications are expressed by a percentage. Percentage of ripple value of capacitor voltage is given by the ratio $\varepsilon = \Delta V_{Ci}/2V_{Ci}$ with a typical value in a conventional converter [14] between

1% and 2%, and percentage of ripple value of inductor currents regarding the current through them is given by $\varepsilon = \Delta I_{Li}/2I_{Li}$ with a typical value between 10% and 20%. For the cascade buck converter, expressions to compute the ripples of inductor currents and capacitor voltages are giving by:

$$\Delta I_{L_i} = \frac{EU^i(1-U)}{L_i f_S} \quad \text{for } i = 1, \cdots, n \tag{1.15a}$$

$$\Delta V_{C_i} = \frac{U^{2n-i}E(1-U)}{Rf_S C_i} \quad \text{for } i = 1, \cdots, \text{n-1} \tag{1.15b}$$

For the ripple of the capacitor voltages is interesting to note a special characteristic of output stage, it is found that ripple of output capacitor is given by $\Delta V_{C_n} = \dfrac{U^n E(1-U)}{8 f_S^2 L_n C_n}$.

Furthermore, it must be guarantee that inductor currents meet this condition for CCM. As previously indicated by the construction of converter, $i_{L_n} > \cdots i_{L_2} > i_{L_1}$; then, it should be fulfilled that in each inductor the conditions $I_{L_i} + \Delta I_{L_i}/2 > 0$ and $I_{L_i} - \Delta I_{L_i}/2 > 0$ are satisfied. For the cascade buck converter, the inductor values that guarantee that CCM condition are given by inequality $L_i > (1-U)R / 2f_S U^{2(n-i)}$ for $i = 1, \cdots, n$.

Cascade Boost Converter

The operation of cascade boost converter with a single switch is based on the principle of quadratic boost converter shown in section 1.4. In this converter, output stage is a conventional converter, meanwhile the first stage is formed by an inductor, a capacitor and a diode. Using the active switch, the boost converters operate in cascade.

Thus, if $n-1$ boost converters, as the one described in the former, are connected by diodes to the active switch, results in a ratio conversion given by $(1-U)^n$, which is shown in Figure 1.13. This topology uses n inductors L_i, n capacitors C_i, a single active switch S and $(2n-1)$ diodes D_i. Nominal duty ratio is given by U, input voltage by E, output voltage by V_O, switching frequency by f_S, and load is modeled by R.

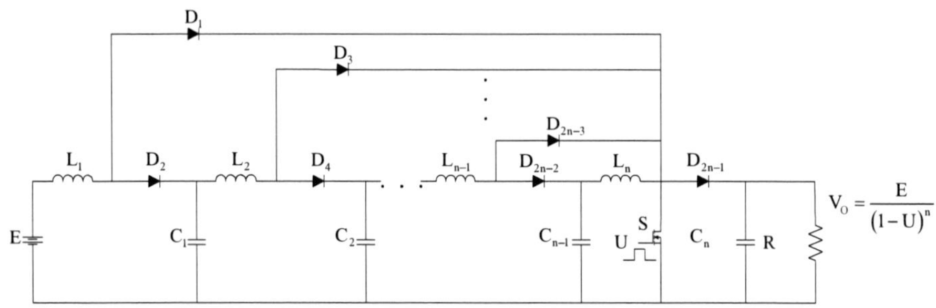

Figure 1.13. Cascade boost converter.

Conversion ratio of converter is derived by assuming that it operates in CCM, and alike ON and OFF times of diodes are synchronized with ON and OFF times of active switch. Figure 1.14 shows when active switch is ON during the part of time of switching function and simultaneously diodes D_1, D_3 to D_{2n-3} are ON; thus, there is a closed trajectory for current. Diodes D_2, D_4 to D_{2n-2} and D_{2n-1} at this time are OFF, thus they do not allow current flow.

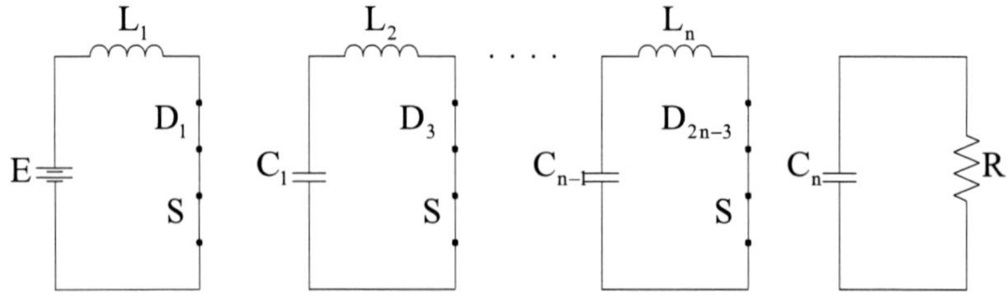

Figure 1.14. Circuit diagram of cascade boost converter with active switch closed.

Later, when active switch S is OFF, during the part of time corresponding to $(1-U)$, simultaneously diodes D_1, D_3 to D_{2n-3} are OFF, and diodes D_2, D_4 to D_{2n-2} and D_{2n-1} turn ON, allowing a new current trajectory, as shown in Figure 1.15.

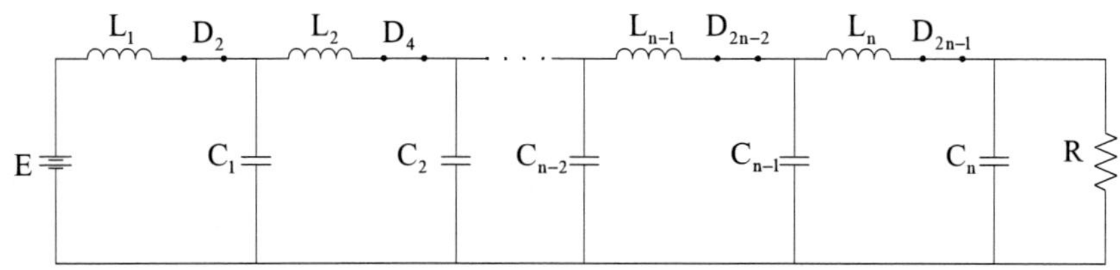

Figure 1.15. Circuit diagram of cascade boost converter with active switch open.

By analyzing the steady state response in the electrical circuits shown in the former, it follows that capacitor voltages are given by $V_{C_i} = E/(1-U)^i$ for $i = 1, \cdots, n$, and inductor currents by $I_{L_i} = I_O/(1-U)^{n+1-i}$ for $i = 1, \cdots, n$ where I_O is the output current. As it can be seen, the capacitor voltages increase as inductor currents decrease throughout the converter. The output voltage is given by $V_o = E/(1-U)^n$.

Expressions to compute the ripples of inductor currents and capacitor voltages of the cascade boost converter are giving by:

$$\Delta I_{L_i} = \frac{EU}{L_i(1-U)^{i-1} f_S} \quad \text{for } i = 1, \cdots, n \tag{1.16a}$$

$$\Delta V_{C_i} = \frac{U^{2n-i} E(1-U)}{Rf_S C_i} \quad \text{for } i = 1, \cdots, n \tag{1.16b}$$

As in cascade buck converter, the study of cascade boost converter is performed in CCM; then, inductor currents must meet $i_{L_1} > i_{L_2} > \ldots i_{L_n} > 0$. Also, the currents in each inductor have to satisfy the inequalities $I_{L_i} + \Delta I_{L_i}/2 > 0$ and $I_{L_i} - \Delta I_{L_i}/2 > 0$. For cascade boost converters, the inductor values for CCM operation are given by $L_i > U(1-U)^{2(n+1-i)} R/2f_S$ for $i = 1, \cdots, n$.

1.6. Switching State Space Models of Cascade Converters

A characteristic of state space modeling is that representations of many systems can be obtained, whether linear, nonlinear, single variable, multivariable, time-invariant or time-variant, by using internal, input and output variables. Furthermore, a compact representation is obtained by a system representation. This matrix representation can be then transformer in output-input system using frequency response methods. Therefore, this method is preferred over that based on input-output signal as transfer function, where information about internal variables can be lost, because some are not observable or controllable states.

Switching modeling of state equations produce a detailed time model where the effects of high frequency could be included, besides to analyze transients effects during ON and OFF times and allows a robust analysis to variation of parameters. Switching systems are analyzed by first linear order equations of state variables of each circuit that results when switches are opened or closed. Finally, when the sum of effects is considered, a matrix representation with the following form is obtained:

$$\dot{x} = F_O x + G_O u + \sum_{i=1}^{m} q_i(x)\left[F_i x + G_i u\right] \tag{1.17}$$

where F_O y G_O are circuit matrix when all switches are OFF, m is the possible combinations of resulting circuits matrices F_i and G_i, $x(t)$ is state vector, $u(t)$ is input vector and q_i the switching function for $i = 1, \cdots, m$.

Cascade Buck Converter Model

Using the electrical circuit diagrams shown in Figures 1.11 and 1.12, and the Kirchhoff's laws, the following expressions are obtained for the cascade buck converter:

Active switch closed:

$$\begin{bmatrix} \dot{i}_{L_1} \\ \dot{i}_{L_2} \\ \vdots \\ \dot{i}_{L_n} \\ \dot{v}_{C_1} \\ \dot{v}_{C_2} \\ \vdots \\ \dot{v}_{C_n} \end{bmatrix} = \begin{bmatrix} 0 & 0 & \cdots & 0 & -1/L_1 & 0 & \cdots & 0 \\ 0 & 0 & \cdots & 0 & 1/L_2 & -1/L_2 & \cdots & 0 \\ \vdots & \vdots & \ddots & 0 & 0 & \vdots & \ddots & 0 \\ 0 & 0 & 0 & 0 & \vdots & 0 & 1/L_n & -1/L_n \\ 1/C_1 & -1/C_1 & 0 & 0 & 0 & 0 & 0 & 0 \\ 0 & 1/C_2 & \ddots & 0 & 0 & 0 & 0 & 0 \\ \vdots & \vdots & \ddots & -1/C_{n-1} & 0 & \vdots & \vdots & 0 \\ 0 & 0 & \cdots & 1/C_n & 0 & 0 & 0 & -1/RC_n \end{bmatrix} \begin{bmatrix} i_{L_1} \\ i_{L_2} \\ \vdots \\ i_{L_n} \\ v_{C_1} \\ v_{C_2} \\ \vdots \\ v_{C_n} \end{bmatrix} + \begin{bmatrix} 1/L_1 \\ 0 \\ \vdots \\ 0 \\ 0 \\ 0 \\ \vdots \\ 0 \end{bmatrix} e \tag{1.18a}$$

Active switch open:

$$\begin{bmatrix} \dot{i}_{L_1} \\ \dot{i}_{L_2} \\ \vdots \\ \dot{i}_{L_n} \\ \dot{v}_{C_1} \\ \dot{v}_{C_2} \\ \vdots \\ \dot{v}_{C_n} \end{bmatrix} = \begin{bmatrix} 0 & 0 & \cdots & 0 & 1/L_1 & 0 & \cdots & 0 \\ 0 & 0 & \cdots & 0 & 0 & 1/L_2 & \cdots & 0 \\ \vdots & \vdots & \ddots & 0 & 0 & \vdots & \ddots & 0 \\ 0 & 0 & 0 & 0 & \vdots & 0 & 0 & 1/L_n \\ 1/C_1 & 0 & 0 & 0 & 0 & 0 & 0 & 0 \\ 0 & 1/C_2 & \ddots & 0 & 0 & 0 & 0 & 0 \\ \vdots & \vdots & \ddots & 0 & 0 & \vdots & \vdots & 0 \\ 0 & 0 & \cdots & 1/C_n & 0 & 0 & 0 & 1/RC_n \end{bmatrix} \begin{bmatrix} i_{L_1} \\ i_{L_2} \\ \vdots \\ i_{L_n} \\ v_{C_1} \\ v_{C_2} \\ \vdots \\ v_{C_n} \end{bmatrix} + \begin{bmatrix} 0 \\ 0 \\ \vdots \\ 0 \\ 0 \\ 0 \\ \vdots \\ 0 \end{bmatrix} e \tag{1.18b}$$

Once the state equations were obtained for each circuit, a switching function represented by q variable is defined. For this converter, q takes a binary value [1, 0]. Using expressions (1.18a) and (1.18b), the switching model of cascade buck converter is obtained as:

$$\begin{bmatrix} \dot{i}_{L_1} \\ \dot{i}_{L_2} \\ \vdots \\ \dot{i}_{L_n} \\ \dot{v}_{C_1} \\ \dot{v}_{C_2} \\ \vdots \\ \dot{v}_{C_n} \end{bmatrix} = \begin{bmatrix} 0 & 0 & \cdots & 0 & -1/L_1 & 0 & \cdots & 0 \\ 0 & 0 & \cdots & 0 & q/L_2 & -1/L_2 & \cdots & 0 \\ \vdots & \vdots & \ddots & 0 & 0 & \vdots & \ddots & 0 \\ 0 & 0 & 0 & 0 & \vdots & 0 & q/L_n & -1/L_n \\ 1/C_1 & -q/C_1 & 0 & 0 & 0 & 0 & 0 & 0 \\ 0 & 1/C_2 & \ddots & 0 & 0 & 0 & 0 & 0 \\ \vdots & \vdots & \ddots & -q/C_{n-1} & 0 & \vdots & \vdots & 0 \\ 0 & 0 & \cdots & 1/C_n & 0 & 0 & 0 & -1/RC_n \end{bmatrix} \begin{bmatrix} i_{L_1} \\ i_{L_2} \\ \vdots \\ i_{L_n} \\ v_{C_1} \\ v_{C_2} \\ \vdots \\ v_{C_n} \end{bmatrix} + \begin{bmatrix} q/L_1 \\ 0 \\ \vdots \\ 0 \\ 0 \\ 0 \\ \vdots \\ 0 \end{bmatrix} e \tag{1.19}$$

This matrix representation has the form $\dot{x} = F(q)x + G(q)e$ in which $x \in \mathbb{R}^{2n}$ are the states, $e(t) \in \mathbb{R}$ the input voltage; F a matrix of dimensions $2n \times 2n$, G a column vector of dimension 2n, and n the number of stages.

Cascade Boost Converter Model

For the boost cascade converter, the circuit diagrams are shown in Figures 1.14 and 1.15. Using again the Kirchhoff's laws, the following expressions are obtained:

Active switch closed:

$$\begin{bmatrix} \dot{i}_{L_1} \\ \dot{i}_{L_2} \\ \vdots \\ \dot{i}_{L_n} \\ \dot{v}_{C_1} \\ \dot{v}_{C_2} \\ \vdots \\ \dot{v}_{C_n} \end{bmatrix} = \begin{bmatrix} 0 & 0 & \cdots & 0 & 0 & 0 & \cdots & 0 \\ 0 & 0 & \cdots & 0 & 1/L_2 & 0 & \cdots & 0 \\ \vdots & \vdots & \ddots & 0 & 0 & \vdots & \ddots & 0 \\ 0 & 0 & 0 & 0 & \vdots & 0 & 1/L_n & 0 \\ 0 & -1/C_1 & 0 & 0 & 0 & 0 & 0 & 0 \\ 0 & 0 & \ddots & 0 & 0 & 0 & 0 & 0 \\ \vdots & \vdots & \ddots & -1/C_{n-1} & 0 & \vdots & \vdots & 0 \\ 0 & 0 & \cdots & 0 & 0 & 0 & 0 & -1/RC_n \end{bmatrix} \begin{bmatrix} i_{L_1} \\ i_{L_2} \\ \vdots \\ i_{L_n} \\ v_{C_1} \\ v_{C_2} \\ \vdots \\ v_{C_n} \end{bmatrix} + \begin{bmatrix} 1/L_1 \\ 0 \\ \vdots \\ 0 \\ 0 \\ 0 \\ \vdots \\ 0 \end{bmatrix} e \tag{1.20a}$$

Active switch open:

$$\begin{bmatrix} \dot{i}_{L_1} \\ \dot{i}_{L_2} \\ \vdots \\ \dot{i}_{L_n} \\ \dot{v}_{C_1} \\ \dot{v}_{C_2} \\ \vdots \\ \dot{v}_{C_n} \end{bmatrix} = \begin{bmatrix} 0 & 0 & \cdots & 0 & -/L_1 & 0 & \cdots & 0 \\ 0 & 0 & \cdots & 0 & 1/L_2 & -1/L_2 & \cdots & 0 \\ \vdots & \vdots & \ddots & 0 & 0 & \vdots & \ddots & 0 \\ 0 & 0 & 0 & 0 & \vdots & 0 & 1/L_n & -1/L_n \\ 1/C_1 & -1/C_1 & 0 & 0 & 0 & 0 & 0 & 0 \\ 0 & 1/C_2 & \ddots & 0 & 0 & 0 & 0 & 0 \\ \vdots & \vdots & \ddots & -1/C_{n-1} & 0 & \vdots & \vdots & 0 \\ 0 & 0 & \cdots & 1/C_n & 0 & 0 & 0 & -1/RC_n \end{bmatrix} \begin{bmatrix} i_{L_1} \\ i_{L_2} \\ \vdots \\ i_{L_n} \\ v_{C_1} \\ v_{C_2} \\ \vdots \\ v_{C_n} \end{bmatrix} + \begin{bmatrix} 1/L_1 \\ 0 \\ \vdots \\ 0 \\ 0 \\ 0 \\ \vdots \\ 0 \end{bmatrix} e \tag{1.20b}$$

For the cascade boost converter, it is possible to define, as in the case of cascade buck converter, a switching function q, which corresponds to active switch ON. Using this function, the switching state space model is given by:

$$\begin{bmatrix} \dot{i}_{L_1} \\ \dot{i}_{L_2} \\ \vdots \\ \dot{i}_{L_n} \\ \dot{v}_{C_1} \\ \dot{v}_{C_2} \\ \vdots \\ \dot{v}_{C_n} \end{bmatrix} = \begin{bmatrix} 0 & 0 & \cdots & 0 & -(1-q)/L_1 & 0 & \cdots & 0 \\ 0 & 0 & \cdots & 0 & 1/L_2 & -(1-q)/L_2 & \cdots & 0 \\ \vdots & \vdots & \ddots & 0 & 0 & \vdots & \ddots & 0 \\ 0 & 0 & 0 & 0 & \vdots & 0 & 1/L_n & -(1-q)/L_n \\ (1-q)/C_1 & -1/C_1 & 0 & 0 & 0 & 0 & 0 & 0 \\ 0 & (1-q)/C_2 & \ddots & 0 & 0 & 0 & 0 & 0 \\ \vdots & \vdots & \ddots & -1/C_{n-1} & 0 & \vdots & \vdots & 0 \\ 0 & 0 & \cdots & (1-q)/C_n & 0 & 0 & 0 & -1/RC_n \end{bmatrix} \begin{bmatrix} i_{L_1} \\ i_{L_2} \\ \vdots \\ i_{L_n} \\ v_{C_1} \\ v_{C_2} \\ \vdots \\ v_{C_n} \end{bmatrix} + \begin{bmatrix} 1/L_1 \\ 0 \\ \vdots \\ 0 \\ 0 \\ 0 \\ \vdots \\ 0 \end{bmatrix} e \qquad (1.21)$$

From the switching state space models the corresponding average models will be obtained in the next section.

1.7. AVERAGE MODELING OF CASCADE CONVERTERS

Average Model of Cascade Buck Converter

Using the average modeling technique describe in section 1.2, the resulting average model of cascade buck converter is giving by:

$$\begin{bmatrix} \dot{\bar{i}}_{L_1} \\ \dot{\bar{i}}_{L_2} \\ \vdots \\ \dot{\bar{i}}_{L_n} \\ \dot{\bar{v}}_{C_1} \\ \dot{\bar{v}}_{C_2} \\ \vdots \\ \dot{\bar{v}}_{C_n} \end{bmatrix} = \begin{bmatrix} 0 & 0 & \cdots & 0 & -1/L_1 & 0 & \cdots & 0 \\ 0 & 0 & \cdots & 0 & \bar{u}/L_2 & -1/L_2 & \cdots & 0 \\ \vdots & \vdots & \ddots & 0 & 0 & \vdots & \ddots & 0 \\ 0 & 0 & 0 & 0 & \vdots & 0 & \bar{u}/L_n & -1/L_n \\ 1/C_1 & -\bar{u}/C_1 & 0 & 0 & 0 & 0 & 0 & 0 \\ 0 & 1/C_2 & \ddots & 0 & 0 & 0 & 0 & 0 \\ \vdots & \vdots & \ddots & -\bar{u}/C_{n-1} & 0 & \vdots & \vdots & 0 \\ 0 & 0 & \cdots & 1/C_n & 0 & 0 & 0 & -1/RC_n \end{bmatrix} \begin{bmatrix} \bar{i}_{L_1} \\ \bar{i}_{L_2} \\ \vdots \\ \bar{i}_{L_n} \\ \bar{v}_{C_1} \\ \bar{v}_{C_2} \\ \vdots \\ \bar{v}_{C_n} \end{bmatrix} + \begin{bmatrix} \bar{u}/L_1 \\ 0 \\ \vdots \\ 0 \\ 0 \\ 0 \\ \vdots \\ 0 \end{bmatrix} \bar{e} \qquad (1.22)$$

This is a nonlinear representation, because matrix F and vector G both depend on control signal $\bar{u}(t) \in \frown$.

Average Model of Cascade Boost Converter

For the cascade boost converter [33], the procedure described at the beginning of this section is applied. Thus, the average cascade boost converter model results in:

$$
\begin{bmatrix} \dot{\bar{i}}_{L_1} \\ \dot{\bar{i}}_{L_2} \\ \vdots \\ \dot{\bar{i}}_{L_n} \\ \dot{\bar{v}}_{C_1} \\ \dot{\bar{v}}_{C_2} \\ \vdots \\ \dot{\bar{v}}_{C_n} \end{bmatrix}
=
\begin{bmatrix}
0 & 0 & \cdots & 0 & -(1-\bar{u})/L_1 & 0 & \cdots & 0 \\
0 & 0 & \cdots & 0 & 1/L_2 & -(1-\bar{u})/L_2 & \cdots & 0 \\
\vdots & \vdots & \ddots & 0 & 0 & \vdots & \ddots & 0 \\
0 & 0 & 0 & 0 & \vdots & 0 & 1/L_n & -(1-\bar{u})/L_n \\
(1-\bar{u})/C_1 & -1/C_1 & 0 & 0 & 0 & 0 & 0 & 0 \\
0 & (1-\bar{u})/C_2 & \ddots & 0 & 0 & 0 & 0 & 0 \\
\vdots & \vdots & \ddots & -1/C_{n-1} & 0 & \vdots & \vdots & 0 \\
0 & 0 & \cdots & (1-\bar{u})/C_n & 0 & 0 & 0 & -1/RC_n
\end{bmatrix}
\begin{bmatrix} \bar{i}_{L_1} \\ \bar{i}_{L_2} \\ \vdots \\ \bar{i}_{L_n} \\ \bar{v}_{C_1} \\ \bar{v}_{C_2} \\ \vdots \\ \bar{v}_{C_n} \end{bmatrix}
+
\begin{bmatrix} 1/L_1 \\ 0 \\ \vdots \\ 0 \\ 0 \\ 0 \\ \vdots \\ 0 \end{bmatrix} \bar{e}
\tag{1.23}
$$

In this model representation, the matrix F depends on control signal $\bar{u}$; and therefore, this is a nonlinear model.

1.8. Linear Average Models for Cascade Converters

As it was indicated in the former section, the average models obtained for cascade buck and boost converters are nonlinear, because the control signal appears as product of some state variables. For simplicity, these systems can be studied using theory of linear systems. When a linear representation for switching converters is obtained, it is possible to use linear systems tools as Laplace transform and frequency domain representations. In addition, it allows using design concepts well defined as gain margin and phase gain, or well-established poles and zeroes interpretations of transfer functions. Linearization process is Taylor series expansion around an operating point or equilibrium, by means of which the system is approximate to one of first order. Therefore, a linearized model describes the behavior of a system in the presence of small disturbances around an operating point.

Average Linear Model for Cascade Buck Converter

To carry out the linearization of average model, using the above procedure, variables are rewritten as a nominal value denoted by an uppercase letter and deviations denoted by a superscript ~; that is: $\bar{u} = U + \tilde{u}$, $\bar{e} = E + \tilde{e}$, $\bar{v}_{C_i} = V_{C_i} + \tilde{v}_{C_i}$ y $\bar{i}_{L_i} = I_{L_i} + \tilde{i}_{L_i}$ for $i = 1, \cdots, n$. The resulting average linear model of cascade buck converter is given by:

$$\begin{bmatrix} \dot{\tilde{i}}_{L_1} \\ \dot{\tilde{i}}_{L_2} \\ \vdots \\ \dot{\tilde{i}}_{L_n} \\ \dot{\tilde{v}}_{C_1} \\ \dot{\tilde{v}}_{C_2} \\ \vdots \\ \dot{\tilde{v}}_{C_n} \end{bmatrix} = \begin{bmatrix} 0 & 0 & \cdots & 0 & -1/L_1 & 0 & \cdots & 0 \\ 0 & 0 & \cdots & 0 & U/L_2 & -1/L_2 & \cdots & 0 \\ \vdots & \vdots & \ddots & 0 & 0 & \vdots & \ddots & 0 \\ 0 & 0 & 0 & 0 & \vdots & 0 & U/L_n & -1/L_n \\ 1/C_1 & -U/C_1 & 0 & 0 & 0 & 0 & 0 & 0 \\ 0 & 1/C_2 & \ddots & 0 & 0 & 0 & 0 & 0 \\ \vdots & \vdots & \ddots & -U/C_{n-1} & 0 & \vdots & \vdots & 0 \\ 0 & 0 & \cdots & 1/C_n & 0 & 0 & 0 & -1/RC_n \end{bmatrix} \begin{bmatrix} \tilde{i}_{L_1} \\ \tilde{i}_{L_2} \\ \vdots \\ \tilde{i}_{L_n} \\ \tilde{v}_{C_1} \\ \tilde{v}_{C_2} \\ \vdots \\ \tilde{v}_{C_n} \end{bmatrix}$$

$$+ \begin{bmatrix} E/L_1 \\ EU/L_2 \\ \ddots \\ EU^{n-1}/L_n \\ -U^{2n-2}E/RC_1 \\ \ddots \\ -U^{n}E/RC_{n-1} \\ 0 \end{bmatrix} \tilde{u} \tag{1.24}$$

where the variations in the input source have been neglected for simplicity.

This expression has the form $\dot{\tilde{x}} = F\tilde{x} + G\tilde{v}$, where $\tilde{x}(t) \in \mathbb{R}^{2n}$ is the state vector and $\tilde{v}(t) \in \mathbb{R}^2$ is the input vector, the matrix F is a matrix of dimensions $2n \times 2n$, vector G of 2n dimension and n the number of stages. As in the nonlinear case, resulting model is similar to that obtained for several buck converters connected in cascade with several active switches [26] when control signal is the same for all active switches.

Average Linear Model of Cascade Boost Converter

The following model is obtained when the linearization is applied to average nonlinear equations:

$$\begin{bmatrix} \dot{\tilde{i}}_{L_1} \\ \dot{\tilde{i}}_{L_2} \\ \vdots \\ \dot{\tilde{i}}_{L_n} \\ \dot{\tilde{v}}_{C_1} \\ \dot{\tilde{v}}_{C_2} \\ \vdots \\ \dot{\tilde{v}}_{C_n} \end{bmatrix} = \begin{bmatrix} 0 & 0 & \cdots & 0 & -(1-U)/L_1 & 0 & \cdots & 0 \\ 0 & 0 & \cdots & 0 & 1/L_2 & -(1-U)/L_2 & \cdots & 0 \\ \vdots & \vdots & \ddots & 0 & 0 & \vdots & \ddots & 0 \\ 0 & 0 & 0 & 0 & \vdots & 0 & 1/L_n & -(1-U)/L_n \\ (1-U)/C_1 & -1/C_1 & 0 & 0 & 0 & 0 & 0 & 0 \\ 0 & (1-U)/C_2 & \ddots & 0 & 0 & 0 & 0 & 0 \\ \vdots & \vdots & \ddots & -1/C_{n-1} & 0 & \vdots & \vdots & 0 \\ 0 & 0 & \cdots & (1-U)/C_n & 0 & 0 & 0 & -1/RC_n \end{bmatrix} \begin{bmatrix} \tilde{i}_{L_1} \\ \tilde{i}_{L_2} \\ \vdots \\ \tilde{i}_{L_n} \\ \tilde{v}_{C_1} \\ \tilde{v}_{C_2} \\ \vdots \\ \tilde{v}_{C_n} \end{bmatrix}$$

$$+ \begin{bmatrix} E/(1-U)L_1 \\ E/(1-U)^2 L_2 \\ \ddots \\ E/(1-U)^n L_n \\ -E/(1-U)^{2n} RC_1 \\ -E/(1-U)^{2n-1} RC_2 \\ \ddots \\ -E/(1-U)^{n+1} RC_n \end{bmatrix} \tilde{u} \qquad (1.25)$$

This expression can be rewritten in the form $\dot{\tilde{x}} = F\tilde{x} + G\tilde{v}$ where $\tilde{x} \in \mathbb{R}^{2n}$ is the state vector and $\tilde{v}(t) \in \mathbb{R}^2$ the input vector. The matrix F is of dimensions $2n$ □ $2n$, vector G of dimension $2n$ and number of converter stages is n.

1.9. SUMMARY OF MODELS

Throughout this chapter, switching, nonlinear and linear models for the different topologies of converters have been developed. These models are for a class of converters that use only an active switch for its operation. Thus, it is very interesting to see that these models represent a whole family of converters that operate with a single switch, as conventional converters or quadratic converters described in sections 1.2 and 1.4, respectively.

To obtain the specific models, it is only necessary to replace in the general model the number of stages. For conventional converters, the resulting expressions are shown in Table 1.1. For quadratic converters, the corresponding models are shown in Table 1.2 for buck converter and in Table 1.3 for boost converter, respectively. These tables allow to verify the models obtained in the last sections of this chapter, correspond to those already used to represent one and two stages; thus, the proposed models are extended to several stages for the case when only an active switch is used to operate a converter.

Table 1.1. Models for conventional converters

Model	Buck converter	Boost converter
Switching state space model	$\begin{bmatrix} \dot{i}_L \\ \dot{v}_C \end{bmatrix} = \begin{bmatrix} 0 & -\frac{1}{L} \\ \frac{1}{C} & -\frac{1}{RC} \end{bmatrix} \begin{bmatrix} i_L \\ v_C \end{bmatrix} + \begin{bmatrix} \frac{q}{L} \\ 0 \end{bmatrix} e$	$\begin{bmatrix} \dot{i}_L \\ \dot{v}_C \end{bmatrix} = \begin{bmatrix} 0 & -\frac{(1-q)}{L} \\ \frac{(1-q)}{C} & -\frac{1}{RC} \end{bmatrix} \begin{bmatrix} i_L \\ v_C \end{bmatrix} + \begin{bmatrix} \frac{1}{L} \\ 0 \end{bmatrix} e$
Average nonlinear state space model	$\begin{bmatrix} \dot{\bar{i}}_L \\ \dot{\bar{v}}_C \end{bmatrix} = \begin{bmatrix} 0 & -\frac{1}{L} \\ \frac{1}{C} & -\frac{1}{RC} \end{bmatrix} \begin{bmatrix} \bar{i}_L \\ \bar{v}_C \end{bmatrix} + \begin{bmatrix} \frac{\bar{u}}{L} \\ 0 \end{bmatrix} \bar{e}$	$\begin{bmatrix} \dot{\bar{i}}_L \\ \dot{\bar{v}}_C \end{bmatrix} = \begin{bmatrix} 0 & -\frac{(1-\bar{u})}{L} \\ -\frac{(1-\bar{u})}{C} & -\frac{1}{RC} \end{bmatrix} \begin{bmatrix} \bar{i}_L \\ \bar{v}_C \end{bmatrix} + \begin{bmatrix} \frac{1}{L} \\ 0 \end{bmatrix} \bar{e}$
Average linear state space model	$\begin{bmatrix} \dot{\tilde{i}}_L \\ \dot{\tilde{v}}_C \end{bmatrix} = \begin{bmatrix} 0 & -\frac{1}{L} \\ \frac{1}{C} & -\frac{1}{RC} \end{bmatrix} \begin{bmatrix} \tilde{i}_L \\ \tilde{v}_C \end{bmatrix} + \begin{bmatrix} \frac{E}{L} \\ 0 \end{bmatrix} \tilde{u}$	$\begin{bmatrix} \dot{\tilde{i}}_L \\ \dot{\tilde{v}}_C \end{bmatrix} = \begin{bmatrix} 0 & -\frac{(1-U)}{L} \\ -\frac{(1-U)}{C} & -\frac{1}{RC} \end{bmatrix} \begin{bmatrix} \tilde{i}_L \\ \tilde{v}_C \end{bmatrix} + \begin{bmatrix} \frac{E}{(1-U)L} \\ \frac{E}{(1-U)^2 RC} \end{bmatrix} \tilde{u}$

Table 1.2. Models for the quadratic buck converter

Switching state space model	$\begin{bmatrix} \dot{i}_{L_1} \\ \dot{i}_{L_2} \\ \dot{v}_{C_1} \\ \dot{v}_{C_2} \end{bmatrix} = \begin{bmatrix} 0 & 0 & -\frac{1}{L_1} & 0 \\ 0 & 0 & \frac{q}{L_2} & -\frac{1}{L_2} \\ \frac{1}{C_1} & -\frac{q}{C_1} & 0 & 0 \\ 0 & \frac{1}{C_2} & 0 & -\frac{1}{RC_2} \end{bmatrix} \begin{bmatrix} i_{L_1} \\ i_{L_2} \\ v_{C_1} \\ v_{C_2} \end{bmatrix} + \begin{bmatrix} \frac{q}{L_1} \\ 0 \\ 0 \\ 0 \end{bmatrix} e$
Average nonlinear state space model	$\begin{bmatrix} \dot{\bar{i}}_{L_1} \\ \dot{\bar{i}}_{L_2} \\ \dot{\bar{v}}_{C_1} \\ \dot{\bar{v}}_{C_2} \end{bmatrix} = \begin{bmatrix} 0 & 0 & -\frac{1}{L_1} & 0 \\ 0 & 0 & \frac{\bar{u}}{L_2} & -\frac{1}{L_2} \\ \frac{1}{C_1} & -\frac{u}{C_1} & 0 & 0 \\ 0 & \frac{1}{C_2} & 0 & -\frac{1}{RC_2} \end{bmatrix} \begin{bmatrix} \bar{i}_{L_1} \\ \bar{i}_{L_2} \\ \bar{v}_{C_1} \\ \bar{v}_{C_2} \end{bmatrix} + \begin{bmatrix} \frac{\bar{u}}{L_1} \\ 0 \\ 0 \\ 0 \end{bmatrix} \bar{e}$
Average linear state space model	$\begin{bmatrix} \dot{\tilde{i}}_{L1} \\ \dot{\tilde{i}}_{L2} \\ \dot{\tilde{v}}_{C1} \\ \dot{\tilde{v}}_{C2} \end{bmatrix} = \begin{bmatrix} 0 & 0 & -\frac{1}{L_1} & 0 \\ 0 & 0 & \frac{U}{L_2} & -\frac{1}{L_2} \\ \frac{1}{C_1} & -\frac{U}{C_1} & 0 & 0 \\ 0 & \frac{1}{C_2} & 0 & -\frac{1}{RC_2} \end{bmatrix} \begin{bmatrix} \tilde{i}_{L1} \\ \tilde{i}_{L2} \\ \tilde{v}_{C1} \\ \tilde{v}_{C2} \end{bmatrix} + \begin{bmatrix} \frac{U}{L_1} \\ \frac{UE}{L_1} \\ -\frac{U^2E}{RC_1} \\ 0 \end{bmatrix} \tilde{u}$

Table 1.3. Models for the quadratic boost converter

Switching state space model	$\begin{bmatrix}\dot{i}_{L1}\\ \dot{i}_{L2}\\ \dot{v}_{C1}\\ \dot{v}_{C2}\end{bmatrix} = \begin{bmatrix}0 & 0 & -\frac{(1-q)}{L_1} & 0\\ 0 & 0 & \frac{1}{L_2} & -\frac{(1-q)}{L_2}\\ \frac{(1-q)}{C_1} & -\frac{1}{C_1} & 0 & 0\\ 0 & \frac{(1-q)}{C_2} & 0 & -\frac{1}{RC_2}\end{bmatrix}\begin{bmatrix}i_{L1}\\ i_{L2}\\ v_{C1}\\ v_{C2}\end{bmatrix} + \begin{bmatrix}\frac{1}{L_1}\\ 0\\ 0\\ 0\end{bmatrix} e$
Average nonlinear state space model	$\begin{bmatrix}\dot{\bar{i}}_{L1}\\ \dot{\bar{i}}_{L2}\\ \dot{\bar{v}}_{C1}\\ \dot{\bar{v}}_{C2}\end{bmatrix} = \begin{bmatrix}0 & 0 & -\frac{(1-\bar{u})}{L_1} & 0\\ 0 & 0 & \frac{1}{L_2} & -\frac{(1-\bar{u})}{L_2}\\ \frac{(1-\bar{u})}{C_1} & -\frac{1}{C_1} & 0 & 0\\ 0 & \frac{(1-\bar{u})}{C_2} & 0 & -\frac{1}{R C_2}\end{bmatrix}\begin{bmatrix}\bar{i}_{L1}\\ \bar{i}_{L2}\\ \bar{v}_{C1}\\ \bar{v}_{C2}\end{bmatrix} + \begin{bmatrix}\frac{1}{L_1}\\ 0\\ 0\\ 0\end{bmatrix} \bar{e}$
Average linear state space model	$\begin{bmatrix}\dot{\tilde{i}}_{L_1}\\ \dot{\tilde{i}}_{L_2}\\ \dot{\tilde{v}}_{C_1}\\ \dot{\tilde{v}}_{C_2}\end{bmatrix} = \begin{bmatrix}0 & 0 & -\frac{(1-U)}{L_1} & 0\\ 0 & 0 & \frac{1}{L_2} & -\frac{(1-U)}{L_2}\\ \frac{(1-U)}{C_1} & -\frac{1}{C_1} & 0 & 0\\ 0 & \frac{(1-U)}{C_2} & 0 & -\frac{1}{RC_2}\end{bmatrix}\begin{bmatrix}\tilde{i}_{L_1}\\ \tilde{i}_{L_2}\\ \tilde{v}_{C_1}\\ \tilde{v}_{C_2}\end{bmatrix} + \begin{bmatrix}\frac{E}{(1-U)^2 L_2}\\ -\frac{E}{(1-U)^4 RC_1}\\ -\frac{E}{(1-U)^4 RC_1}\\ -\frac{E}{(1-U)^3 RC_2}\end{bmatrix} \tilde{u}$

CONCLUSION

Nowadays, there is a need of converters with wide range of voltage conversion ratios, and as consequence, the design of regulators that provide good characteristics and performance. Thus, it is very important to have models that describe the dynamic behavior of converters. With these models, different control strategies can be used in order to design a regulator for specific applications.

An approach to the problem to step up or down output voltage is to use conventional converters operating at extremely low or at extremely high duty ratios; however, multiple problems arise as the turn ON and turn OFF times of the switching elements are very small. Other solutions include the use of transformers or converters connected in parallel (interleaved converters). A drawback of these topologies is the presence of severe peaks in the switching devices, because of inductor linkage between windings as well as the need of complex control circuits. The use of several conventional converters connected is cascade offer a possible solution, being its main drawback the complex control scheme. To avoid the use of many active switches, cascade converters with a single active switch are proposed. These topologies show efficiencies slightly higher than those of several active switches but their control system is much easier to design. For cascade converters with a single switch, relationship of ripple of inductor currents, ripple of capacitor voltages and inductor values for continuous conduction mode are developed. Using state space modeling techniques, switching, average nonlinear and linear or small-signal models are shown.

REFERENCES

[1] J. Bebic, R. Walling, K. O'Brien, B. Kroposki, "The sun also rises," *IEEE Power Energy Mag.*, vol. 7, no. 3, p.p. 45–54, 2009.

[2] F. Orechini, A. Santiangeli and A. Delli'Era, "A Technological Solution for an Everywhere Energy Supply With Sun, Hydrogen and Fuel Cells," *ASME J. Fuel Cell Sci. Technol*, vol. 3, no.1, pp 75-82, 2005.

[3] P. C. Sernia and G. R.Walker, "Cascaded dc-dc converter connection of photovoltaic modules", *IEEE Trans. Power Electron.*, Vol. 19, No. 4, pp. 1130-1139, 2004.

[4] M. Veerachary, T. Senjyu, and K. Uetazo, "Maximum power point tracking of coupled inductor interleaved boost converter supplied PV system", *IEE Proc. Elect. Power Appl.*, vol. 150, No. 1, pp. 71-80, 2003.

[5] F. Blaabjerg, R. Teodorescu, M. Liserre, A.V. Timbus, "Overview of Control and Grid Synchronization for Distributed Power Generation Systems", *IEEE Trans. Ind. Electron.*, vol. 53 , no. 5 , p.p. 1398–1409, 2006.

[6] M. H. Todorovic, L. Palma, P. N. Enjeti, "Design of a wide input range DC-DC converter with a robust power control scheme suitable for fuel cell power conversion," *IEEE Trans. Ind. Electron.*, vol. 55, no. 3, pp. 1247-1255, 2008.

[7] L.-S. Yang, T.-J. Liang, J.-F. Chen "Transformerless DC-DC converters with high step-up voltage gain," *IEEE Trans. Ind. Electron.*, vol. 56, no. 8, pp. 3144-3152, 2009.

[8] S.-K.Changchien, T.-J. Liang, J.-F. Chen, L.-S. Yang, "Novel high step-up DC-DC converter for fuel cell energy conversion system," *IEEE Trans. Ind. Electron.*, vol. 57, no. 6, pp. 2007-2017, 2010.

[9] A. Shahin, M. Hinaje, J.-P. Martin, S. Pierfederici, S. Rael, B. Daval, "High voltage ratio DC-DC converter for fuel-cell applications", *IEEE Trans. Ind. Electron.,* vol. 57, no. 12, pp. 3944-3955, 2010.

[10] X. Zou, P.I. Wong, F.C. Lee, A.Q. Huang, "Investigation of candidate VRM topologies for future microprocessors", *IEEE Trans. Power Electron.,* vol. 15, no. 6, p.p. 1172-1182, 2000.

[11] S. Benerjee, and G. C. Verghese, Nonlinear Phenomena in Power Electronics, Attractor, Bifurcations, Chaos and Nonlinear Control. *IEEE* Press, 2001.

[12] R. W. Erickson and D. Maksimovic, Fundamentals of Power Electronics. Second Edition, Kluwer Academic Publishers, 2001.

[13] N. Mohan, T. M. Undeland, and W. P. Robbins, *Power Electronics: Converters, Applications and Design*. New York, John Wiley & Sons, Third Edition, 2003.

[14] P. T. Krein, *Elements of Power Electronics*, First Edition, Oxford University Press, 1998.

[15] J. G. Kassakian, M. F. Schlecht, and G. C. Verghese, *Principles of Power Electronics.* Addison-Wesley Publishing Company, 1991.

[16] M. H. Rashid, Power Electronics, Circuit, Devices and Applications. Englewood Cliffs, N.J., Prentice Hall, Second Edition, 1993.

[17] M. Verachary, "General rules for signal flow graph modeling and analysis of dc-dc converters", *IEEE Trans. Aerosp. Electron. Syst.,* vol. 40, no. 1, pp. 259-271, 2004.

[18] M. Veerachary, "Modelling and analysis of cascade step-down converters", *IEE Proc. Electr. Power Appl.*, vol. 152, no. 1, pp. 41-50, 2005.

[19] M. Veerachary, "Signal flow graph model of multi-state boost DC-DC converters", *IEE Proc. Electr. Power Appl.*, vol. 151, no. 5, pp. 583-589, 2004.

[20] J. Y. Zhu and B. Lehman, "Control loop design for two-stage dc-dc converters with low voltage/high current output", *IEEE Trans. Power Electron.,* vol. 20, No. 1, pp. 44-55, 2005.

[21] T. Qian and B. Lehman, "Buck half-bridge input-series two stage converters," *IET Power Electron.*, vol.3, no. 6, p.p. 965-978, 2010.

[22] P. Xu, J. Wei, F.C. Lee, "Investigation of candidates topologies for 12 V VRM," *IEEE Trans. Power Electron.*, vol. 18, no. 1, p.p. 686-691, 2002.

[23] P. Xu, F.C. Lee, "Multiphase coupled-buck converter –a novel high efficient 12 V voltage regulator module," *IEEE Trans. Power Electron.*, vol. 18, no. 1, p.p. 74-82, 2003.

[24] Y. Panov and M.M. Javanovic, "Design considerations for 12 V/1.5V, 50 A, voltage regulators modules," *IEEE Trans. Power Electron.*, vol. 16, no. 6, p.p. 778-783, 2001.

[25] H. Matsuo and K. Harada, "The cascade connection of switching regulators", *IEEE Trans. Ind. Appl.*, vol. 12, no. 2, pp. 192-198, March/April 1976.

[26] J. A. Morales-Saldaña, E. E. Carbajal-Gutierrez and J. Leyva-Ramos, "Modeling of switch-mode DC-DC cascade converters", *IEEE Trans. Aerosp. Electron. Syst.*, vol. 38, no. 1, pp. 295-299, 2002.

[27] D. Maksimovic and S. Cuk, "Switching converters with wide dc conversion range", *IEEE Trans. Power Electron.*, vol. 6, no. 1, pp. 151-157, 1991.

[28] F. L. Luo and H. Ye, Advanced DC/DC Converters, CRC Press, 2003.

[29] F. L. Luo and H. Ye, "Positive output cascade boost converters", *IEE Proc. Electr. Power Appl.*, vol. 131, no. 5, pp. 590-606, 2004.

[30] F. L. Luo and H. Ye, "Positive output super-lift converters", *IEEE Trans. Power Electron.,* vol. 18, no. 1, 2003.

[31] J. A. Morales-Saldaña, J. Leyva-Ramos, E. E. Carbajal-Gutierrez and M. G. Ortiz-Lopez, "A current–mode control scheme for a quadratic buck converter with a single switch", *IEEE Trans. Power Electron.*, vol. 23, no.1, p.p. 485-490, 2008.

[32] J. A. Morales-Saldaña, R. Galarza-Quirino, E. E. Carbajal-Gutierrez, J. Leyva-Ramos and M. G. Ortiz-Lopez, "Multiloop controller design for a quadratic boost converter", *IET Proc. Electr. Power Appl.*, vol. I, No. 3, pp. 362-367, 2007.

[33] J. Leyva-Ramos, M. G. Ortiz-Lopez, L. H. Diaz-Saldierna, J. A. Morales-Saldana, "Switching regulator using a quadratic boost converter for wide DC conversion ratios" *IET Power Electron.*, vol. 2, no. 5, p.p. 605-613, 2009.

In: Hydroelectricity and Power Electronics
Editor: Lucille Burke
ISBN: 978-1-63321-728-7

Chapter 6

AlGaN/GaN BASED-POWER MOSFET WITH HIGHLY RESISTIVE UNDOPED BUFFER LAYER

Jae-Hoon Lee*

Discrete Development Team, System LSI,
Samsung Electronics Co., Ltd, Giheung, Korea

ABSTRACT

High performance normally-off GaN-based metal oxide semiconductor field effect transistor (MOSFET) was fabricated. The buffer layer of the MOSFET was grown by varying the growth temperature to control the size of nucleation sites which results in extremely high buffer resistance(>10^{12} Ω/sq). The fabricated small-area MOSFET exhibited excellent normally-off device characteristics, such as threshold voltage of 2 V, maximum drain current of 253 mA/mm, on-off current ratio of 5.5×10^7, destructive breakdown voltage of 830 V, and leakage current of 0.7 μA/mm at V_{DS} of 600 V. The corresponding values of the large-area MOSFET with multi-finger pattern were 0.6 V, 6 A, 1.3×10^7, 670 V, and 50 μA (0.28 μA/mm).

* Corresponding author: Email: jaehoon.lee@samsung.com.

1. INTRODUCTION

Group III–nitride semiconductors and their ternary solid solutions are very promising as the candidates for both short wavelength optoelectronics and power electronic devices because of the intrinsic advantages of materials such as direct bandgap and high electron peak velocity [1]-[5]. The wide bandgap nature of the AlGaN/GaN heterostructure leads to high device breakdown voltage, which is very important in high power switching applications such as converters and inverters, including solid-state drives for heavy motors, pulsed power for electric vehicles or ships [6]. Also the AlGaN/GaN-based power device exhibits very high switching speed, originated from the high electron velocity of the material, which can simplify the power conversion system and reduce the size of the system. Although remarkable progress has been achieved in the GaN-based high power and high frequency electronic device, some performance related issues still need an immediate solution.

It is well known that the usual undoped GaN layer grown on normal growth condition generally exhibits n-type conductivity due to residual donor impurities such as nitrogen vacancies and oxygen atoms which cannot be used as a buffer layer of the AlGaN/GaN-based heterostructure field effect transistors (HFETs) [7-10]. Therefore, the semi-insulating GaN layer is required to avoid the parallel conduction and to minimize the off-state leakage current. The semi-insulating GaN layer was successfully obtained by doping compensating acceptors such as C, Fe, or Mg, even though some problems such as redistribution of dopant atoms toward the channel region may exist [11], [12]. These impurities can also exhibit strong memory effect in epitaxial growth thereby precluding abrupt transitions in doping. To avoid these problems, introduction of threading dislocations or other defects during the growth was suggested to obtain the semi-insulating GaN layer. However, the device fabricated on this semi-insulating buffer layer suffers from the trapping/detrapping of electrons due to the degraded crystalline quality which occasionally results in a severe current collapse in AlGaN/GaN HFETs [13]. To minimize the trapping effect, therefore, it is necessary to grow a semi-insulating GaN layer while maintaining the crystalline quality as high as possible.

Slow-acting trap state trap electron and create a charged layer that depletes the channel in the high field region between the gate and drain. Since the time constants of these states range from seconds to micro-seconds, the 2DEG channel cannot be fully modulated during a large signal RF operation, and the RF current collapse eventually decreases the output power [14, 15]. Such slow-acting trap states are associated with the surface state and threading dislocations in the AlGaN barrier layer, as confirmed by scanning capacitance microscopy [16]. The surface-states are believed to be responsible for the leakage current of the AlGaN/GaN HFETs and occasionally cause a severe current collapse when the HFETs operate under high power and high frequency conditions. To reduce surface leakage current, various methods such as surface passivation, surface treatments with either chemicals or plasma, and post annealing, have been investigated. Dielectric films, such as SiN_x, SiO_2, HfO_2 and Al_2O_3, have been used as surface passivation layers as well as gate insulators for the AlGaN/GaN HFETs [17]-[19]. A modified AlGaN/AlN/GaN HFETs structured with an AlN interfacial layer was also proposed to reduce the current collapse [20].

In this work, we suggest a simple method of growing a highly resistive GaN buffer layer grown on sapphire substrate by controlling the size of the nucleation sites. Higher device performances for the metal oxide semiconductor field effect transistors (MOSFETs), fabricated on the AlGaN/GaN heterostructure with this proposed buffer layer, are also demonstrated.

2. Fabrication of AlGaN/GaN-Based MOSFET

The AlGaN/GaN heterostructure investigated in this work was grown by metal organic chemical vapor deposition (MOCVD) on 4 inch (0001) c-plane sapphire substrates which consists of 30 nm-thick low temperature-grown initial GaN layer, 3 μm-thick highly resistive undoped GaN buffer layer, and 25 nm-thick AlGaN barrier. Differently from the normal growth, the undoped GaN buffer layer of the heterostructure was first grown at relatively low temperature of 950 °C for 2 min. and the temperature was elevated to the normal growth temperature of 1020 °C to complete the remaining growth. The time for the temperature ramping was 2 min. and no growth interruption was applied. The Al content in the AlGaN barrier is 30%, determined by high-resolution X-ray diffraction. The mobility and the density of the two dimensional electron gas (2DEG) formed at the AlGaN/GaN heterointerface were 1600 $cm^2/V\cdot s$ and $8\times10^{12}/cm^2$, respectively. For the device fabrication, the active region of the device was defined by inductively coupled plasma (ICP) reactive ion etching using a BCl_3/Cl_2 gas mixture. The 25 nm-thick AlGaN layer in the gate region was fully recessed by using the same gas mixture. A 25 nm-thick Al_2O_3 gate insulator layer was then deposited on the recessed GaN surface by plasma assisted atomic layer deposition (PAALD) at 450 °C, subsequently annealed at 600°C for 30 min. in N_2 ambient to recover plasma damage and improve the dielectric quality. After opening contact holes, Ti/Al/Ni/Au metal layer for ohmic contact was deposited and followed by rapid thermal annealing at 850 °C for 30 s in N_2 ambient. The specific contact resistance of 2×10^{-5} $\Omega\cdot cm^{-2}$ was obtained for the annealed sample using transmission line measurements (TLM). After depositing Ni/Au for the gate metal, Si_3N_4 inter-dielectric layer with thickness of 800 nm was deposited to cover the entire surface of the device. Ti/Al pad metals were finally deposited to connect the gate and the source/drain region. A schematic cross-sectional and transmission electron microscope (TEM) image of the fabricated MOSFET is shown in Figure 1[21].

3. Growth of Semi-Insulation GaN Buffer Layer

Five different undoped GaN samples were grown on a 16 nm-thick LT-GaN buffer at temperatures of 950, 980, 1000, 1020, and 1050° for 40 min. The surface morphology of the layers was measured by both atomic force microscopy (AFM) and scanning electron microscopy (SEM). The electrical and optical properties were measured by a Hall-effect measurement and photoluminescence(PL), while the crystal quality of the films was determined by high-resolution X-ray diffraction (HR-XRD). Figure 2 shows SEM and AFM images of the as-grown and annealed LT-buffer GaN at different temperatures. The as-grown LT-buffer layer is shown in Figure 1(a), while three subsequent LT-buffer layers were annealed at (b) 950°, (c) 1020°, and (d) 1050°. The SEM and AFM images show a gradual increase in the polycrystalline island size with an increasing annealing temperature. The values of the root mean square(rms) of the surface roughness were (a) 2.4 nm, (b) 7.1 nm, (c) 22 nm, and (d) 24 nm, respectively. With a relatively thin LT-buffer layer thickness (16 nm), Ga species become more mobile on a sapphire surface with an increasing temperature and form larger polycrystalline islands [22]. Figure 3 shows the HRXRD diffraction data for the same samples presented in Figure2. The FWHM values for the X-ray rocking curve were (a) 13786 arcsec, (b) 8420 arcsec, (c) 2714 arcsec, and (d) 2310 arcsec, respectively. A broader rocking curve was observed for the as-deposited sample and sample ramped up to 950° based on 4 min. However, when the annealing temperature was increased above 1020°, the single-crystalline GaN quality was improved. Figure 4 shows the PL spectra at 10 K for the samples presented in Figure 2. As mentioned above, no near band edge peak at 365 nm was observed for the as-grown sample and sample annealed up to 950º based on 4 min, yet when the annealing temperature was increased, the intensity of the near band edge emission peak also increased. As such, these results indicate a change in the crystallinity of the LT-GaN layer, from amorphous to single crystal, during the ramping process. Figure 5 shows the results of the sheet resistance and electron mobility for undoped GaN films grown at various temperatures for 40 min. With a decrease in the growth temperature, the value of the sheet resistance greatly increased from ~ 10^3 to ~ 10^6 Ω/sq, while the electron mobility also significantly decreased from 250 to 4 cm^2/Vs. Free carrier concentration was reduced from 5×10^{16} cm^{-3} (~5×10^{12} cm^{-2}) for the 1.7 μm-GaN films grown on sapphire substrate at 1020º to 1×10^{16} cm^{-3} (~1×10^{12} cm^{-2}) for the sample grown at 950º. When the ramping temperature was slowly increased from 550 to 950º based on 4 min, smaller grain sizes and a higher nuclei density were observed, as shown in Figure 2(b). Since the recrystallization sites are not large enough and misorientation of the nuclei occurs in the direction of the a or c axis with a slower and lower ramping temperature, an HT-GaN layer grown under these conditions contains high density dislocations and stacking faults [23]. Thus, the high sheet resistance was believed to be due to generation of various point defect centers at the coalescence boundaries as a result of a small nucleation size in initial growth stage. In contrast, larger grain sizes and a lower nuclei density, as shown in Figs. 2 (c) and (d), allowed more room for individual islands to develop before coalescence. Although the larger intermediate islands resulted in a rougher surface morphology, the increased average volume of defect-free columnar domains and lower density of coalescence boundaries eventually produced improved structural and electrical properties [23], as shown in Figure 5. When using a typical single-step growth method, a high sheet resistance was obtained at a relatively low growth temperature near 950º, yet the surface morphology of the resultant film was shown to be 3-D in fashion, as shown in Figure 6. However, when the epilayer growth temperature was increased above 1000º, a smooth surface resulted from the high thermal energy. Thus, to obtain a smooth surface with a high sheet resistance, a special two-step growth was implemented to resolve the problem encountered with single-step growth.

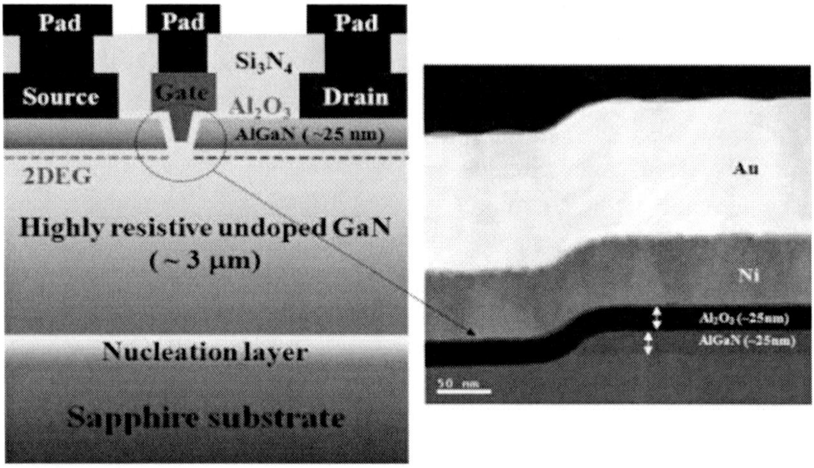

Figure 1. A schematic cross-sectional and TEM image of the fabricated MOSFET.

Figure 2. SEM and AFM images of as-grown buffer and annealed LT-buffer at different temperature: (a) initial buffer at 550°C (RMS: 2.4 nm), (b) 950 °C (RMS: 7.1 nm), (c) 1020 °C (RMS: 22 nm), (d) 1050 °C (RMS: 24 nm).

Figure 7 shows the two-step growth procedure. First, 16 nm LT-GaN was annealed at 950° with a ramping time of 4 min, then the GaN was grown at this temperature for 1 min. Second, the growth temperature was increased to 1020° with a ramping time of 2 min and the GaN layer finally grown at 1020° for 40 min to improve the crystal quality. The film grown using this sequence turned out to be semi-insulating (> 10^{12} Ω/sq) with a mirror-like surface morphology (rms: 8 nm), as shown by the inset in Figure 7. The observation of high resistance is believed due to this initial sequence formed the deep trap levels in between the band gap [24,25].

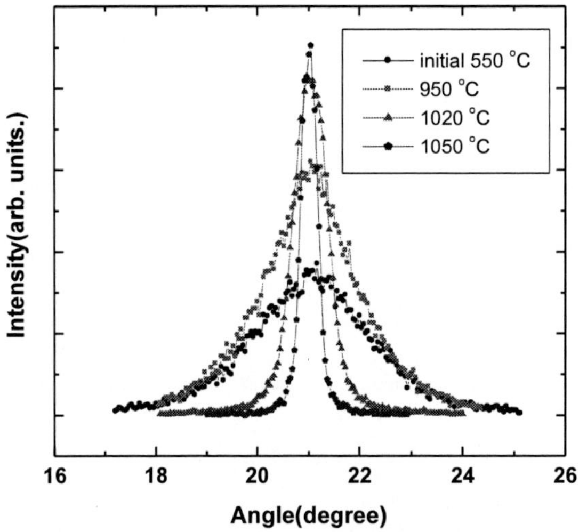

Figure 3. HRXRD diffraction data for as-grown and LT-buffer GaN at different annealing temperatures: initial buffer at 550°, 950°, 1020°, and 1050°.

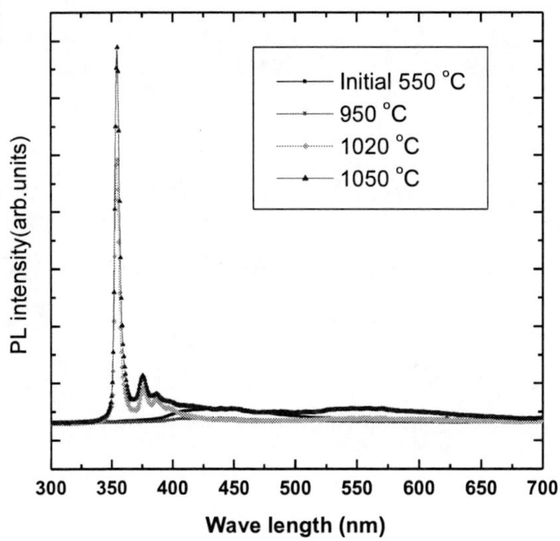

Figure 4. PL spectra at 10 K for as-grown and LT-buffer GaN at different annealing temperatures: initial buffer at 550°, 950°, 1020°, and 1050°.

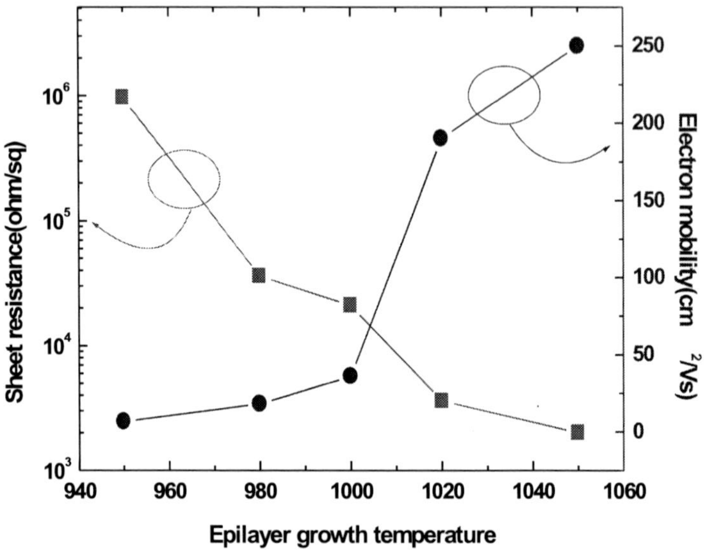

Figure 5. Sheet resistance and electron mobility of undoped GaN films grown at various temperatures: 950°, 980°, 1000°, 1020°, and 1050°.

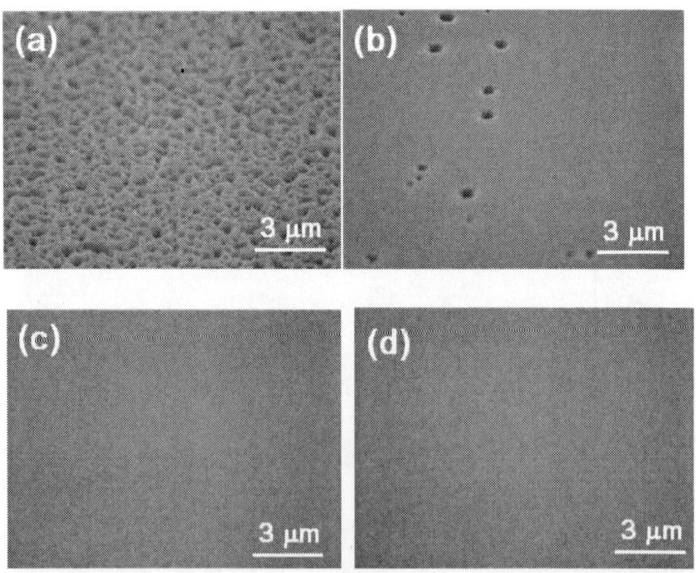

Figure 6. SEM photographs for undoped GaN films grown by single step ramping at various temperatures: (a) 950 °C, (b) 980 °C, (c) 1020 °C, (d) 1050 °C.

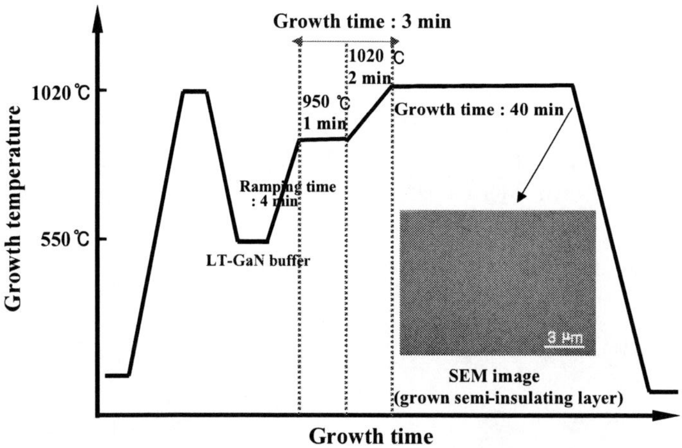

Figure 7. Proposed two-step growth procedure. Inset shows SEM photographs of semi-insulating GaN grown based on two-step growth.

To investigate the growth procedure of highly resistive GaN, figure 8 shows SEM and AFM images of a sample(a) grown at 1020° for 3 min based on typical one-step growth and a sample(b) grown at 950° for 1 min, then at an increasing temperature up to 1020° for 2 min based on the proposed two-step growth. The rms values for the surface roughness were 36 nm for the one-step growth and 2.67 nm for the two-step growth. When compared to the one-step growth, the surface roughness of the two-step growth was improved due to the small and numerous grain sizes developed during the initial growth at a relatively low temperature. This implies that smaller grain sizes were formed and the columnar growth mode along the c direction was dominant during the initial low temperature ramping and low temperature growth, thereafter, misorientation of the nuclei in the direction of the a or c axis was promoted due to the GaN growth during the temperature ramping from 950 to 1020°.

Figure 9 shows cross-sectional TEM images under g=0002 two-beam conditions proved sensitive to screw threading dislocation centers for samples grown (a) by one-step process at 1020° and (b) by two-step process from 950 to 1020°. The screw threading dislocation density near the interface between the low-temperature buffer and overlayer was roughly estimated to be $\sim 10^9$ cm^{-2} for the one-step growth and $\sim 10^{10}$ cm^{-2} for the two-step growth, respectively. Thus, the observed threading dislocation density of the two GaN samples was not high enough to capture all of the calculated average carrier density of $\sim 10^{12}$ cm^{-2}. However, with a consideration of multiple electron traps around a dislocation center and/or some other electron traps incarnated into the GaN layers by the presence of the interfacial threading dislocations[24,25], we can speculate roughly that since the one-step sample with a threading dislocation density of $\sim 10^9$ cm^{-2} indicates an average carrier density of $\sim 10^{12}$ cm^{-2} resulting in a sheet resistance of 10^6 Ω/sq, the two-step sample with a dislocation density of $\sim 10^{10}$ cm^{-2} would also result in a high resistance over 10^9 Ω/sq. Well-oriented investigations for the origin of probable electron traps should be made to systematically produce this highly resistive GaN layer.

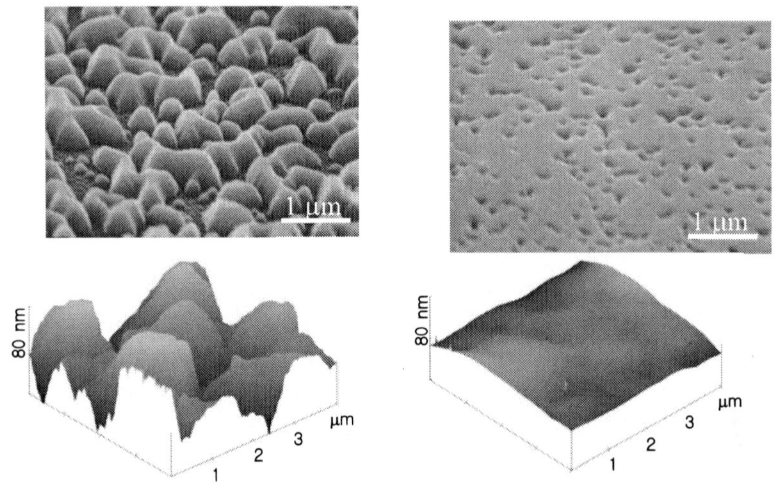

(a) (b)

Figure 8. SEM and AFM images of as-grown and LT-buffer GaN grown at different annealing temperatures: (a) sample grown at 1020° for 3 min based on typical one-step growth and (b) sample grown at 950° for 1 min and at increasing temperature up to 1020° for 2 min based on proposed two-step growth.

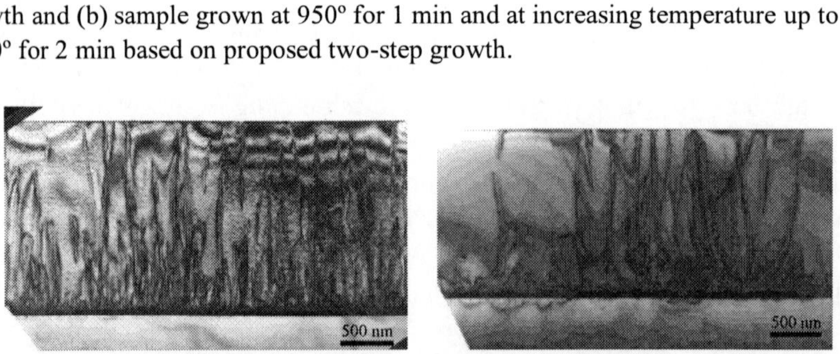

(a) (b)

Figure 9. Cross-sectional TEM images of typical one-step sample (b) and special two-step sample under 0002 two-beam.

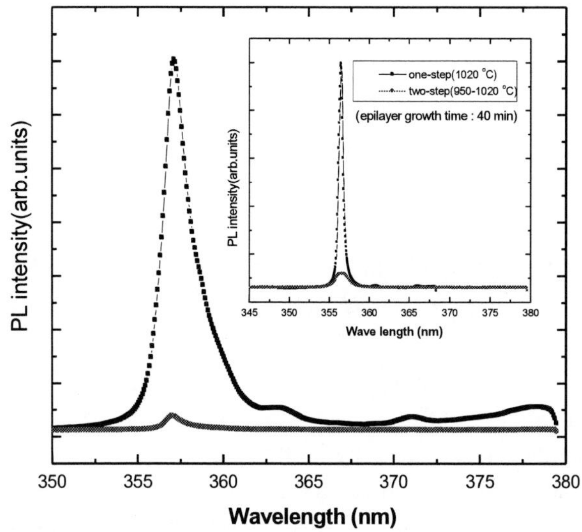

Figure 10. PL spectra for samples grown based on one-step growth and two-step growth for 3 min. Inset shows PL spectra for both samples grown for 40 min.

Figure 10 shows the PL spectra at 10 K for the samples grown based on one-step growth and two-step growth for 3 min, as shown in Figure 8. The PL intensity of the one-step sample increased approximately 17 times compared to that of the two-step sample, indicating that the two-step procedure generated nonradiative recombination centers. In the case of epilayer growth for 40 min, the PL intensity of the one-step sample also increased approximately 10 times compared to that of the two-step sample, as shown by the insert in Fig 10. As such, these results show that the semi-insulating GaN was acquired from compensating free carriers resulting from the many acceptor-like point defects(associated with Ga vacancy) generated at the relatively low ramping and growth temperature and change in growth temperature from 950 to 1020° based on 3 min for epilayer growth[26, 27].

To evaluate the trapping effect in the buffer layer, the pulsed I_{DS} -V_{DS} measurement, as shown in Figure 11, was performed for the normally-on small-sized GaN HFET device which has been fabricated on the same GaN buffer layer as the GaN power MOSFET. The gate length and width of the device are 2 and 140 μm, respectively. To measure the gate- and the drain-lag, the bias voltage changed from V_{DS} = 0.0V/V_{GS} = 0.0V to V_{DS} = 0.0V/V_{GS} = - 7.0 V and from V_{DS} = 20V/V_{GS} = - 2.0V to V_{DS} = 20V/V_{GS} = - 7.0V, respectively[28]. The duration and separation of the pulse were 0.5 μs and 0.5

ms, respectively. The pulse measurement exhibits that the current collapse related to the gate- and the drain-lag is insignificant for the device. Negligible drain-lag observed for the device probably confirms that the trap levels in the proposed semi-insulating GaN buffer layer are very deep and hence the trapping/detrapping of electrons through the traps is not active.

4. Characteristics of the Normally-off AlGaN/GaN-MOSFET

Figure 12 shows the current–voltage (I-V) characteristics of the fully recessed AlGaN/GaN MOSFET with the GaN buffer layer grown by the proposed sequence. The gate length, the gate width, and the gate-to-drain distance of device were 2 and 140, and 20 μm, respectively, and the total size of the device was 500 × 500 μm^2. The device with effective area of 150 × 200 μm^2 (except the contact pad area) exhibits a maximum drain current (I_{dmax}) of 253 mA/mm at a drain voltage of 10 V and the specific on state-resistance is 10 $m\Omega \cdot cm^2$, as shown in Figure 12 (a). The transfer curve (I_{DS}-V_{GS}) of the device at a fixed V_{DS} of 8 V is shown in Figure 12 (b), showing a threshold voltage of 2 V, extrapolated from the I-V characteristics in the linear region (at V_{DS} of 2 V), and a very high on/off drain-current ratio (I_{ON}/I_{OFF}) of 5.5×10^7. These excellent device performances are because the very high buffer resistance achieved by the proposed growth sequence greatly decreases the off-state current and, in addition, the high quality ALD-Al_2O_3 gate insulator ensures low gate leakage current of 2×10^{-6} A/mm (at V_{GS} = 30 V). Figure 12 (d) demonstrates the excellent off-state characteristics with very low leakage current of 0.7 μA/mm at 600 V and high destructive breakdown voltage of 870 V.

Figure 13 shows the device performance for the large-sized MOSFET with the multi-finger pattern for the high power application. The large device has the same device configurations as the small device, except that the gate width and the total size of the device were increased to 180 mm and 3000 × 3000 μm^2, respectively. I_{dmax} is as high as 6 A at V_{DS} of 4 V an d at V_{GS} = 5 V as shown in Figure 13 (a), which corresponds to the normalized value of 30 mA/mm, 5 times lower compared to that of the small-sized device (150 mA/mm at V_{GS}=5 V). This is mainly due to the decreased channel mobility at

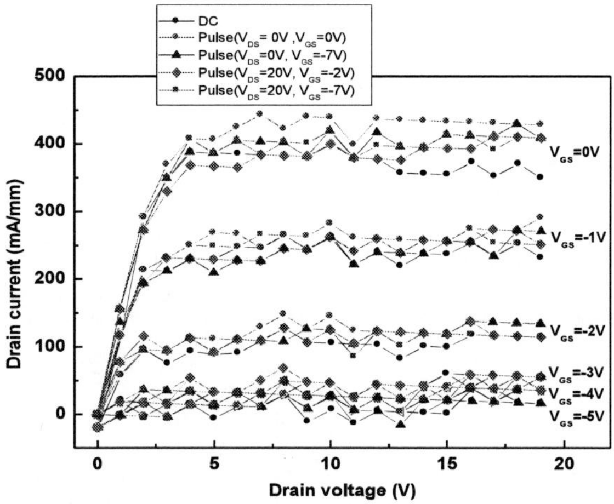

Figure11. Pulsed I_{DS} -V_{DS} characteristics for the normally-on small-sized GaN HFET.

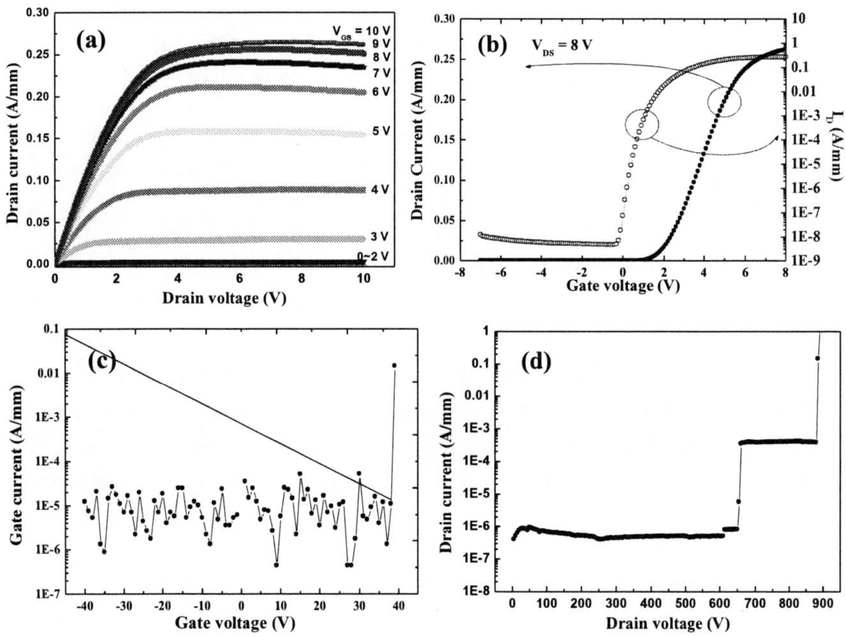

Figure 12. I-V characteristics of the fabricated unit patterned morally-off MOSFETs with 25 nm-recessed gate; (a) drain current, (b) transfer characteristics, (c) gate leakage current, and (d) breakdown voltage.

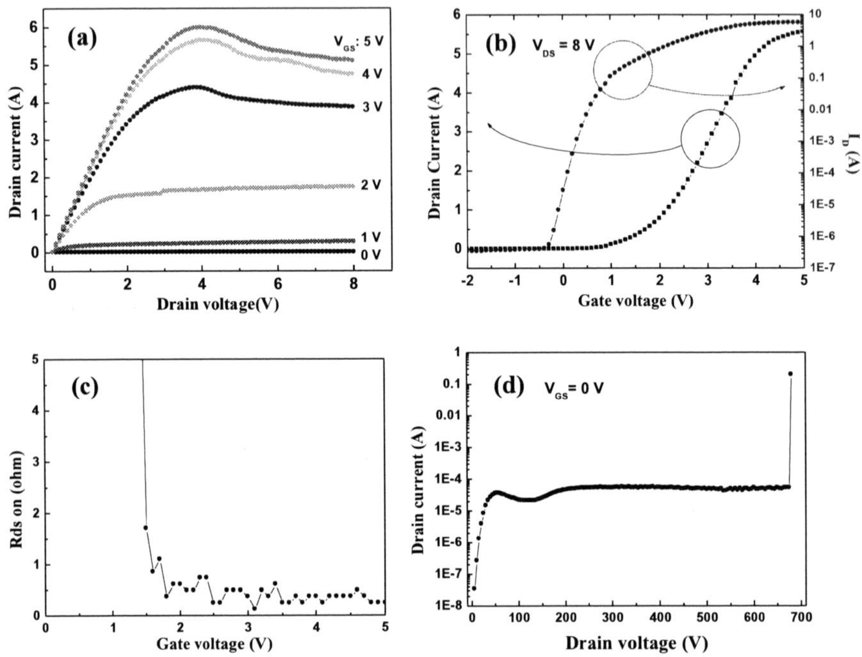

Figure 13. I-V characteristics of the large-sized morally-off MOSFET with the multi-finger pattern; (a) drain current, (b) transfer characteristics, (c) on-resistance, and (d) breakdown voltage.

elevated device temperature under high current operation and also the significant current crowding effect caused by high series resistance in such a large area device. The threshold voltage and I_{ON}/I_{OFF} ratio of 0.6 V and 1.3×10^7 were observed at V_{DS} of 8 V as shown in Figure 13 (b), which are slightly decreased compared to the values obtained for the small-sized devices. The reason of the lower threshold voltage for the large-sized device is may be due to the non-uniform 2DEG density, resulted from the non-uniform recess etching of AlGaN barrier over large surface area. The on-resistance ($R_{DS\ (on)}$) of the large-sized device with dimension of $3000 \times 3000\ \mu m^2$ is 0.3 Ω (27 m$\Omega \cdot cm^2$) at I_{DS} = 1 A and V_{GS} = 5 V as shown in Figure 13 (c). In spite of large chip size, however, the reverse breakdown voltage and the leakage current (measured at V_{DS} of 600 V) were not significantly degraded, showing 670 V and 50 μA

Conclusion

The growth of obtaining highly resistive GaN buffer layer was proposed, which varies the growth temperature during the layer growth to control the sizes of the nucleation sites. The layers grown by the proposed sequence were shown to be semi-insulating with extremely high sheet resistance ($>10^{12}$ Ω/sq), even though any additional compensation impurities were doped. The successful normally-off operation was achieved from the GaN MOSFET fabricated on the AlGaN/GaN heterostructure with this buffer layer. The large-sized MOSFET with multi-finger pattern exhibits a threshold voltage of 0.6 V, maximum drain current of 6A, I_{ON}/I_{OFF} of $\sim10^7$, breakdown voltage of 670 V, and leakage current of 50 μA at V_{DS} of 600 V.

References

[1] Wu, Y. F., Kapolnek, D., Ibbetson, J. P., Parikh, P., Keller, B. P. & Mishra, U. K. (2001). "Very-high power density AlGaN/GaN HEMT's," *IEEE Electron Device Lett.*, vol. *48*, 586 - 590.

[2] Nakamura, S., Senoh, M., Nagahama, S., Iwasa, N., Yamada, T., Matsushita, T., Sugimoto, Y. & Kiyoku, H. (1997). "High-Power, Long-Lifetime InGaN Multi-Quantum-Well-Structure Laser Diodes," *Jpn. J. Appl. Phys.*, vol. *36*, L1059-L1061.

[3] Uemoto, Y., Hikita, M., Ueno, H., Matsuo, H., Ishida, H., Yanagihara, M., Ueda, T., Tanaka, T. & Ueda, D. (2007). "Gate injection transistor (GIT)-A normally-off AlGaN/GaN power transistor using conductivity modulation," *IEEE Trans. Electron Devices*, vol. *54*, no.12, 3393-3399.

[4] Lee, J. H., Lee, D. Y., Oh, B. W. & Lee, J. H. (2010). "Comparison of InGaN-based LEDs grown on conventional sapphire and cone-shape-patterned sapphire substrate," *IEEE Trans. Electron Devices*, vol. *57*, no. 1, 157-163.

[5] Ambacher, O., Smart, J., Shealy, J. R., Weimann, N. G., Chu, K., Murphy, M., Schaff, W. J., Eastman, L. F., Dimitrov, R., Wittmer, L., Stutzmann, M., Rieger, W. & Hilsenbeck, J. (1999). "Two-dimensional electron gases induced by spontaneous and piezoelectric polarization charges in N and Ga-face AlGaN/GaN heterostructures," *J. Appl. Phys.*, vol. *85*, no. 6, 3222 - 3233.

[6] Iwakami, S., Machia, O., Yoshimichi, I., Ryohei, B., Masataka, Y., Toshihiro, E., Nobuo, K., Hirokazu, G. & Akio, I. (2007). "Evaluation of AlGaN/GaN Heterostructure Field-Effect Transistors on Si Substrate in Power Factor Correction Circuit," *Jap. J. Appl. Phys.*, Vol. *46*, No. 29, L721-L723.

[7] Look, D. C., Reynolds, D. C., Jones, R. L., Kim, W., Aktas, O., Botchkarev, A., Salvador, A. & Morkoc, H. (1997). "Electrical and optical properties of semi-insulating GaN", *Mater. Sci. Eng. B44*, 423.

[8] Bougrioua, Z., Moerman, I., Sharma, N., Wallis, R. H., Cheyns, J., Jacobs, K., Thrush, E. J., Considine, L., Beanland, R., Farvacque, J. L. & Humphreys, C. (2001). "Material optimisation for AlGaN/GaN HFET applications", *J. Cryst. Growth*, vol. *230*, 573.

[9] Briot, O., Alexis, J. P., Sanchez, S., Gil, B. & Aulombard, L. (1997). "Influence of the VIII molar ratio on the structural and electronic properties of MOVPE grown GaN", *Sol. St. Electr.* vol. *41*, 315.

[10] Sugiura, L., Itaya, K., Nishio, J., Fujimoto, H. & Kokubun, Y. (1997). "Effects of thermal treatment of low temperature GaN buffer layer on the quality of sequent GaN layer", *J. Appl. Phys*., vol. *82*, 4877.

[11] Ohba, Y. & Hatano, A. (1994). "A study on strong memory effects for Mg doping in GaN metalorganic chemical vapor deposition," *J. Cryst. Growth*, vol. *145*, 214.

[12] Heikman, S., Keller, S., DenBaars, S. P. & Mishra, U. K. (2002). "Growth of Fe doped semi-insulating GaN by metalorganic chemical vapor deposition," *Appl. Phys. Lett.,* vol. *81*, no. 3, 439.

[13] Binari, S. C. et al. (2001). "Trapping effects and microwave power performance in AlGaN/GaN HEMTs," *IEEE Trans. Electron Devices*, vol. *48*, 465.

[14] Jong-Wook Kim, Jae-Seung Lee, Jin-Ho Shin, Jae-Hoon Lee, Sung-Ho Hahm, Jung-Hee Lee, Chang-Seok Kim, Jae-Eung Oh & Moo-Whan Shin, (2001). "Influence of pinhole type defect in AlGaN on rf performance of "AlGaNH/GaN HFETs grown by MOCVD," *Phys. Stat. Sol.* (a), vol. *188*, 267.

[15] Green, B. M., Chu, K. K., Chumbes, M., Smart, J. A., Shealy, J. R. & Eastman, L. F. (2000). "The effect of surface passivation on the microwave characteristics of undoped AlGaN/GaN HEMTs", *IEEE Electron Device Lett.*, vol. *21*, 268.

[16] Smith, K. V., Yu, E. T., Redwing, J. M. & Boutros, K. S. (2000)."Local electronic properties of AlGaN/GaN heterostructures probed by scanning capacitance microscopy", *J. Electron. Mater.*, vol. *29*, 274.

[17] Hasegawa, H. & Oyama, S. (Jul. 2002). "Mechanism of anomalous current transport in n-type GaN Schottky contacts," *J. Vac. Sci. Technol. B,* vol. *20*, no. 4, 1647–1655.

[18] Derluyn, J., Boeykens, S., Cheng, K., Vandersmissen, R., Das, J., Ruythooren, W., Degroote, S., Leys, M. R., Germain, M. & Borghs, G. (Sep. 2005). "Improvement of AlGaN/GaN high electron mobility transistor structures by in situ deposition of a Si_3N_4 surface layer," *J. Appl. Phys*., vol. *98*, no. 5, 054 501-1–054 501-5.

[19] Onojima, N., Higashiwaki, M., Suda, J., Kimoto, T., Mimura, T. & Matsui, T. (Feb. 2007). "Reduction in potential barrier height of AlGaN/GaN heterostructures by SiN passivation," *J. Appl. Phys*., vol. *101*, no. 4, 043 703-1–043 703-6.

[20] Lee, J. S., Kim, J. W., Lee, J. H., Kim, C. S., Oh, J. E., Shin, M. W. & Lee, J. H. (2003). "Reduction of current collapse in A lGaN/GaN HFETs using AlN interfacial layer", *Electron Lett.*, vol. *39*, no. 9, 750 -751.

[21] Lee, J. H., Jeong, J. H. & Lee, H. J. (2012). "Normally off GaN power MOSFET grown on sapphire substrate with highly resistive undoped buffer layer", *IEEE Electron Device Lett.*, vol. *33*, 1429 -1431.

[22] Ramer, J. C., Zheng, K., Kranenberg, C. F., Banas, M. & Hersee, S. D. (1996). *Mater. Res. Soc. Symp. Proc., 395*, 225.

[23] Sugiura, L., Itaya, K., Nishio, J., Fujimoto, H. & Kokubun, Y. (1997). "Effects of thermal treatment of low temperature GaN buffer layers on the quality of subsequent GaN layers", *J. Appl. Phys., 82*, 4877.

[24] Desnica, U. V. & Pavlovic, M., Fang, Z.-Q. & Look, D. C. (2002). "Thermoelectric effect spectroscopy of deep levels in semi-insulating GaN", *J. Appl. Phys., 92*, 4126.

[25] Oila, J., Saarinen, K., Wickenden, A. E., Koleske, D. D., Henry, R. L. & Twigg, M. E. (2003). "Ga vacancies and grain boundaries in GaN", *Appl. Phys. Lett., 82*, 1021.

[26] Elsner, J. et al. (1998). "Deep acceptors trapped at threading-edge dislocations in GaN," *Phys. Rev. B*, *Condens. Matter*, vol. *58*, no. 19, 12 571.

[27] Lee, S. M., Belkhir, M. A., Zha, X. Y. & Lee, Y. H. (2000). "Electronic structures of GaN edge dislocations," *Phys. Rev. B*, vol. *61*, 16033.

[28] Heikman, S., Keller, S., DenBaars, S. P. & Mishra, U. K. (2002). "Growth of Fe doped semi-insulating GaN by metalorganic chemical vapor deposition," *Appl. Phys. Lett.*, vol. *81*, no. 3, 439.

In: Hydroelectricity and Power Electronics
Editor: Lucille Burke
ISBN: 978-1-63321-728-7

Chapter 7

Parallel Three-Phase Back-to-Back Converters

***Nady Rocha*[1], *Euzeli Cipriano dos Santos, Jr.*[2,*]**
***and Cursino Brandão Jacobina*[3]**
[1]Federal University of Joao Pessoa, Joao Pessoa, Brazil
[2]Indiana University-Purdue University Indianapolis, Indiana, US
[3]Federal University of Campina Grande, Campina Grande, Brazil

Abstract

This chapter proposes bidirectional three-phase to three-phase back-to-back converters, which includes their generalization with n units in parallel. Such converters can be employed to interface high power renewable energy sources (e.g., wind turbines) to the utility grid. Although the proposed converters are characterized by parallel connection without use of transformers, their control system permits canceling the circulating current among the units in parallel. In addition to proposing new configurations, this chapter presents a comprehensive model of the topologies, and a suitable control strategy for the system with regulation at both sides of the converter. Compared to the conventional three-phase to three-phase back-to-back converter, the proposed topologies allow the following advantages: i) reduction on the current and consequently power ratings of the power switches, ii) reduced harmonic distortion at both sides of converter, and iii) lower dc-link capacitor losses, which has a positive impact on the lifespan of the power system. Additionally, the semiconductor

[*]E-mail address: eudossan@iupui.edu

losses can be reduced when the proposed converter is used. Simulated and experimental results are also presented for validation of the theoretical expectations.

Keywords: Paralleled Converter, Circulating Current, Power Converter

1. Introduction

In the modern society, the utility power grid is supposed to guarantee load management, demand side management, as well as to use the market price of electricity and forecasting of energy (e.g., based on wind and solar renewable sources) in order to optimize the whole electrical distribution system. An optimized power system is expected to have the following characteristics: high efficiency, high availability, good quality of service and high level of security, which leads to the concepts of distributed generation and smart-grid [1]. Those objectives can be obtained in the new power distribution systems through advances in control, communication and information technology, as well as extensive use of power electronics. Indeed, an increased portion of electrical energy coming from renewable sources is possible due to power electronics converters.

Since policy and regulatory initiatives have guaranteed further use of renewable energy sources, the amount of energy processed by converters have reached power limits beyond the devices available in the market. Either parallel converters or switches in parallel have been employed to deal with such a restriction. However, only the connection of converters in parallel allows to improve the power capability, reliability, and efficiency of the energy conversion stage. Usually the operation of converters in parallel requires a transformer for isolation. Weight, size, and cost associated with the transformer may make such solution undesirable. When an isolation transformer is not employed, the reduction of circulating currents among different converter stages is an important objective in the system design.

The main advantages of the parallel connection of converters are summarized below [2], [3], [4] and [5]:

- Reduced switch stress, due to the division of power flow between the converters.
- Increased redundancy of the system because: *(i)* the management of power between the converters and *(ii)* the fault-tolerant capability.

- Lower harmonic distortion of input and output converters, with use of multiple carriers PWM.
- Reduction of harmonic components of the dc-link capacitor current, hence reducing the capacitor stress and increasing its lifespan.
- Possibility of reducing the semiconductor power losses in comparison with the conventional topology.

However, as mentioned earlier, without isolation transformer a circulating current may appear among the converters connected in parallel. This circulating current may cause [6], [7], [8], [9], [10] and [11]: *(i)* distorted current waveforms, *(ii)* unbalanced power flow distribution among the converters, *(iii)* power losses increase, *(iv)* passive component saturation.

To avoid circulating current among the units in parallel, different techniques have been proposed in the technical literature, e.g. :*(i)* either isolated voltages sources or transformers at the ac side [12], [13]; *(ii)* high zero sequence impedance due to the reactors between the phases [14], [11], [15] and [16]; *(iii)* common mode inductors [17], [4], [11] and [18]; and *(iv)* control techniques, such as: modeling and control of the circulating current [19], [8], [9] and [20], discontinuous space vector modulation [7], [11] and [21], one cycle control [22], synchronized control [23].

This chapter proposes parallel three-phase to three-phase back-to-back converters, as presented in Figs. 1(a) and 1(b). Fig. 1(c) shows a conventional three-phase back-to-back converter, which is considered here for comparison purposes. In addition to proposing new configurations, this chapter presents a comprehensive model of the topologies, a suitable control strategy for the system with regulation at both sides of the converter, as well as its generalization. Compared to the conventional three-phase to three-phase back-to-back converter [see Fig. 1(c)], the proposed topologies allow the following advantages: reduction on the current and consequently power ratings of the power switches, ii) reduced harmonic distortion at both sides of converter, and iii) lower dc-link capacitor losses, which has a positive impact on the lifespan of the power system. Additionally, the semiconductor losses can be reduced when the proposed converter is used. Simulated and experimental results are also presented for validation of the theoretical expectations.

The topology in Fig. 1(a) is named as T2B since it has twelve legs and two dc-links, while T1B for the configuration in Fig. 1(b) stands for twelve legs and one dc-link. The potential applications for the proposed converters are: *(i)* uninterruptible power supply (UPS), *(ii)* universal active power filters, *(iii)* three-phase and six-phase motor drive systems, and *(iv)* distributed generation system, notably wind generation.

2. System Model

2.1. Topology T2B

The model of configuration T2B [Fig. 1(a)] is presented in Fig. 2. For converters 1 and 3, the following set of equations can be derived

$$e_{gn} = z_{1n} i_{1n} + v_{1n0_1} - v_{g0_1} \tag{1}$$

$$e_{gn} = z_{3n} i_{3n} + v_{3n0_2} - v_{g0_2} \tag{2}$$

with

$$v_{g0_1} = \frac{1}{3}(v_{1a0_1} + v_{1b0_1} + v_{1c0_1}) \tag{3}$$

$$v_{g0_2} = \frac{1}{3}(v_{3a0_2} + v_{3b0_2} + v_{3c0_2}) \tag{4}$$

where $n = a, b, c$, $z_{1n} = r_{1n} + l_{1n} p$, $z_{3n} = r_{3n} + l_{3n} p$, r_{1n} and r_{3n} are resistance of the inductors L_{1n}, L_{3n}, respectively; l_{1n} and l_{3n} are the inductances of the inductors L_{1n} and L_{3n}, respectively; v_{g0_1} is the voltage between the star-center of the grid (g) and the midpoint dc-link 1 (0_1); v_{g0_2} is the voltage between the star-center of the grid and the midpoint of the dc-link 2 (0_2); v_{1n0_1} are the pole voltages of the converter 1; v_{3n0_2} are the pole voltages of the converter 3; i_{1n} are the input currents of the converter 1; and i_{3n} are the input currents of the converter 3.

The following equations are related to converters 2 and 4,

$$e_{ln} = -z_{2n} i_{2n} + v_{2n0_1} - v_{l0_1} \tag{5}$$

$$e_{ln} = -z_{4n} i_{4n} + v_{4n0_2} - v_{l0_2} \tag{6}$$

with

$$v_{l0_1} = \frac{1}{3}(v_{2a0_1} + v_{2b0_1} + v_{2c0_1}) \tag{7}$$

$$v_{l0_2} = \frac{1}{3}(v_{4a0_2} + v_{4b0_2} + v_{4c0_2}) \tag{8}$$

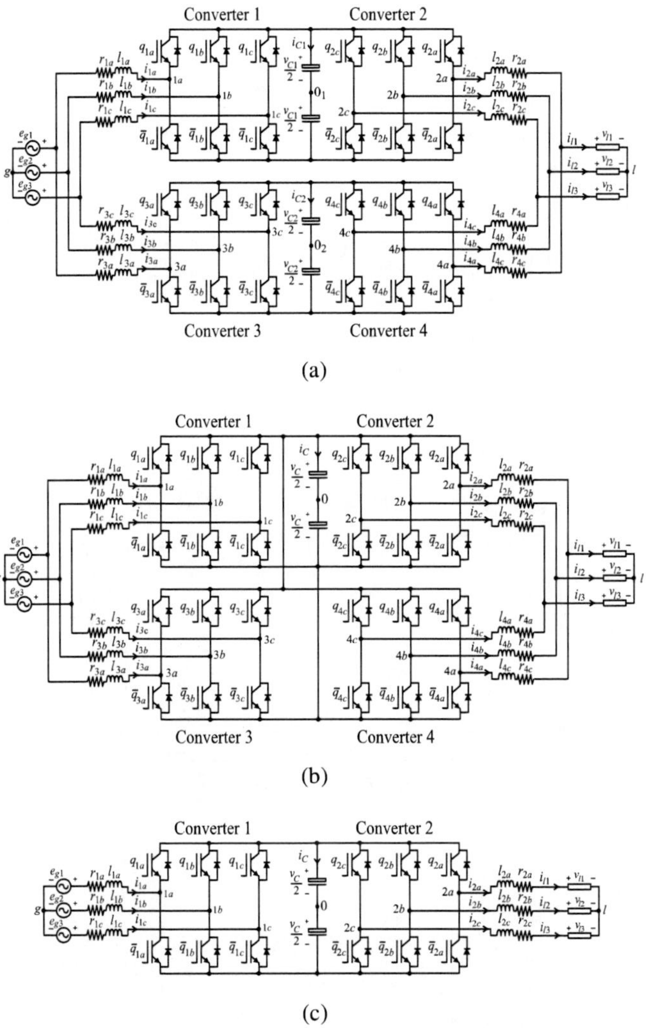

Figure 1. Three-phase back-to-back converters. (a) Proposed topology with twelve legs and two dc-links (T2B). (b) Proposed topology with twelve legs and one dc-link (T1B). (c) Conventional topology with six legs.

where $z_{2n} = r_{2n} + l_{2n}p$ and $z_{4n} = r_{4n} + l_{4n}p$, r_{2n} and r_{4n} represent the resistance of the inductors L_{2n} and L_{4n}, respectively; l_{2n} and l_{4n} represent the inductances of the inductors L_{2n} and L_{4n}, respectively; v_{2n0_1} and v_{4n0_2}

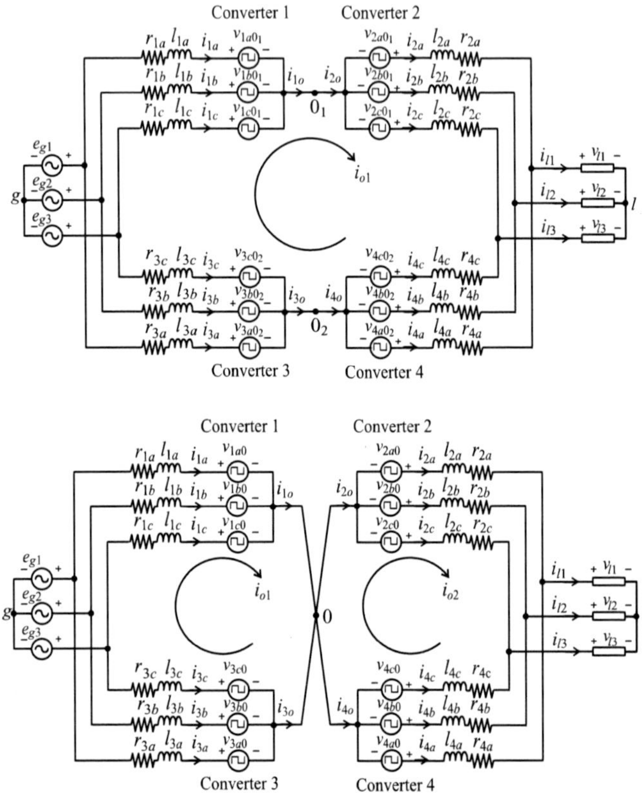

Figure 2. Model of the proposed topologies: (a) configuration T2B and (b) configuration T1B.

represent the pole voltages of the converters 2 and 4, respectively; i_{2n} and i_{4n} are the output currents of the converters 2 and 4, respectively.

Due to the parallel connection there is a circulation loop between both back-to-back converters, which can be modeled by the following equations:

$$z_{1a}i_{1a} + z_{2n}i_{2n} - z_{4n}i_{4n} - z_{3a}i_{3a} = -v_{1a0_1} + v_{2n0_1} - v_{4n0_2} + v_{3a0_2} \quad (9)$$

$$z_{1b}i_{1b} + z_{2n}i_{2n} - z_{4n}i_{4n} - z_{3b}i_{3b} = -v_{1b0_1} + v_{2n0_1} - v_{4n0_2} + v_{3b0_2} \quad (10)$$

$$z_{1c}i_{1c} + z_{2n}i_{2n} - z_{4n}i_{4n} - z_{3c}i_{3c} = -v_{1c0_1} + v_{2n0_1} - v_{4n0_2} + v_{3c0_2} \quad (11)$$

From (9)-(11), the circulating voltage can be defined by

$$v_{o1}=\sum_{n=a}^{c} i_{1n}z_{1n}+\sum_{n=a}^{c} i_{2n}z_{2n}-\sum_{n=a}^{c} i_{4n}z_{4n}-\sum_{n=a}^{c} i_{3n}z_{3n} \tag{12}$$

with

$$v_{o1}=-\sum_{n=a}^{c} v_{1n0_1}+\sum_{n=a}^{c} v_{2n0_1}+\sum_{n=a}^{c} v_{3n0_2}-\sum_{n=a}^{c} v_{4n0_2}. \tag{13}$$

From the input and output currents of the converters (i_{1a}, i_{1b}, i_{2a},..., i_{4b} and i_{4c}) the circulating currents in each converter are defined by:

$$i_{ko}=\frac{1}{\sqrt{3}}\left(i_{ka}+i_{kb}+i_{kc}\right) \tag{14}$$

with $k = 1, 2, 3, 4$. Notice that it is possible to write the equations of the circulating currents of each converter according to a single circulating current (i_{o1}) as following

$$i_{o1}=i_{1o}=i_{2o}=-i_{3o}=-i_{4o} \tag{15}$$

From *odq* transform, considering the balanced system, with $z_{1n} = z_{3n} = z_1$ and $z_{2n} = z_{4n} = z_2$, and making $v_{1n} = v_{1n0_1} - v_{g0_1}$, $v_{2n} = v_{2n0_1} - v_{l0_1}$, $v_{3n} = v_{3n0_2} - v_{g0_2}$ and $v_{4n} = v_{4n0_2} - v_{l0_2}$, from (72), (1)-(2) and (5)-(6), the model of the topology T2B is written as follows:

$$\bar{e}_{gdq} = \bar{z}_1\bar{i}_{1dq} + \bar{v}_{1dq} \tag{16}$$
$$\bar{e}_{gdq} = \bar{z}_1\bar{i}_{3dq} + \bar{v}_{3dq} \tag{17}$$
$$\bar{e}_{ldq} = -\bar{z}_2\bar{i}_{2dq} + \bar{v}_{2dq} \tag{18}$$
$$\bar{e}_{ldq} = -\bar{z}_2\bar{i}_{4dq} + \bar{v}_{4dq} \tag{19}$$
$$v_{o1} = \sqrt{3}\left(2z_1 + 2z_2\right) i_{o1} \tag{20}$$

with $\bar{z}_1 = \begin{bmatrix} z_1 & 0 \\ 0 & z_1 \end{bmatrix}$ and $\bar{z}_2 = \begin{bmatrix} z_2 & 0 \\ 0 & z_2 \end{bmatrix}$.

From (16) and (17), the following relationships can be found for the *dq* grid currents (i_{gd} and i_{gq}):

$$\bar{v}_{gdq} = \frac{\bar{v}_{1dq}+\bar{v}_{3dq}}{2} = \bar{e}_{gdq} - \frac{\bar{z}_1}{2}\bar{i}_{gdq} \tag{21}$$

From (18) and (19) the *dq* load voltages (e_{ld} and e_{lq}) can be calculated by:

$$\bar{v}_{ldq} = \frac{\bar{v}_{2dq}+\bar{v}_{4dq}}{2} = \bar{e}_{ldq} + \frac{\bar{z}_2}{2}\bar{i}_{ldq} \tag{22}$$

It is worth mentioning that the voltages $v_{gd} = (v_{1d} + v_{3d})/2$ and $v_{gq} = (v_{1q} + v_{3q})/2$ define the grid currents i_{gd} and i_{gq}, respectively [see equation (21)], while the voltages $v_{ld} = (v_{2d} + v_{4d})/2$ and $v_{lq} = (v_{2q} + v_{4q})/2$ define the load voltages e_{ld} and e_{lq}, respectively, as shown in (22).

Assuming that $z_g = z_1/2$ and $z_l = z_2/2$, where z_g and z_l are the equivalent inductance of the topology T2B, the system model simplified by

$$\bar{v}_{gdq} = \bar{e}_{gdq} - \bar{z}_g \bar{i}_{gdq} \tag{23}$$

$$\bar{v}_{ldq} = \bar{e}_{ldq} + \bar{z}_l \bar{i}_{ldq} \tag{24}$$

2.2. Topology T1B

The model of topology T1B shown in Fig. 1(b) is presented in Fig. 2(b). The model of converters 1 and 3 is defined by:

$$e_{gn} = z_{1n} i_{1n} + v_{1n0} - v_{g0} \tag{25}$$

$$e_{gn} = z_{3n} i_{3n} + v_{3n0} - v_{g0} \tag{26}$$

with

$$v_{g0} = \frac{1}{6}\left(v_{1a0} + v_{1b0} + v_{1c0} + v_{3a0} + v_{3b0} + v_{3c0}\right) \tag{27}$$

where v_{g0} is the voltage between star-center of the grid (g) and the midpoint of the dc-link (0), v_{1n0} are the pole voltages of the converter 1, v_{3n0} are the pole voltages of the converter 3.

The model of converters 2 and 4 are defined in turn by:

$$v_{ln} = -z_{2n} i_{2n} + v_{2n0} - v_{l0} \tag{28}$$

$$v_{ln} = -z_{4n} i_{4n} + v_{4n0} - v_{l0} \tag{29}$$

with

$$v_{l0} = \frac{1}{6}\left(v_{2a0} + v_{2b0} + v_{2c0} + v_{4a0} + v_{4b0} + v_{4c0}\right) \tag{30}$$

where v_{l0} is the voltage between star-center of the load (l) and the midpoint of the dc-link (0), v_{2n0} are the pole voltages of the converter 2, v_{4n0} are the pole voltages of the converter 4.

Unlikely the configuration T2B with only one circulating current, the topology T1B presents two circulation loops between the converter units as highlighted in Fig. 2(b). Indeed, due to a single dc-link on this configuration, a path is created at the grid side allowing a circulation current between the converters 1 and 3. By analogy another mesh is created at the load side, between the converters 2 and 4. For the converters on the grid side, it yields

$$z_{1n}i_{1n} - z_{3n}i_{3n} + v_{1n0} - v_{3n0} = 0 \tag{31}$$

From (31) the circulating voltage v_{o1} at the grid side is defined by

$$v_{o1} = \sum_{n=a}^{c} i_{1n}z_{1n} - \sum_{n=a}^{c} i_{3n}z_{3n} \tag{32}$$

with

$$v_{o1} = -v_{1a0} - v_{1b0} - v_{1c0} + v_{3a0} + v_{3b0} + v_{3c0}. \tag{33}$$

Similarly, at the load side, it turns out that

$$z_{2n}i_{2n} - z_{4n}i_{4n} - v_{2n0} + v_{4n0} = 0 \tag{34}$$

From (34) the circulating voltage (v_{o2}) is:

$$v_{o2} = \sum_{n=a}^{c} z_{2n}i_{2n} - \sum_{n=a}^{c} z_{4n}i_{4n} \tag{35}$$

with

$$v_{o2} = v_{2a0} + v_{2b0} + v_{2c0} - v_{4a0} - v_{4b0} - v_{4c0} \tag{36}$$

The circulating currents are defined by (14), which leads to

$$i_{o1} = i_{1o} = -i_{3o} \tag{37}$$

$$i_{o2} = i_{2o} = -i_{4o}. \tag{38}$$

Considering a balanced system, $z_g = z_1/2$ and $z_l = z_2/2$, the *odq* model of the topology T1B can be summarized by the following set of equations:

$$\bar{v}_{gdq} = \bar{e}_{gdq} - \bar{z}_g\bar{i}_{gdq} \tag{39}$$

$$\bar{v}_{ldq} = \bar{e}_{ldq} + \bar{z}_l\bar{i}_{ldq} \tag{40}$$

$$v_{o1} = 4\sqrt{3}z_g i_{o1} \tag{41}$$

$$v_{o2} = 4\sqrt{3}z_g i_{o2}. \tag{42}$$

From this model, it is evident that the voltages v_{o1} and v_{o2} will be used to regulated the circulating currents i_{o1} and i_{o2}, while the voltages v_{gdq} and v_{ldq} will be used to control grid currents and load voltages, respectively.

3. PWM Strategy

3.1. Configuration T2B

The pole voltages v_{1n0_1}, v_{2n0_1}, v_{3n0_2} and v_{4n0_2}, with $n = a, b, c$, can be written as a function of conduction states of the power switches, for example for converter 1, is:

$$v_{1n0_1} = (2s_{q1n} - 1)\frac{v_{c_1}}{2} \tag{43}$$

where v_{c_1} is the dc-link voltage and s_{q1n} is the conduction states of the power switches q_{1n}.

Considering that v_{kd}^*, v_{kq}^* and v_{o1}^* represent the reference voltages obtained from the controllers (see Section 4),

$$v_{kd}^* = \sqrt{\frac{2}{3}}\left(v_{ka0_u}^* - \frac{v_{kb0_u}^*}{2} - \frac{v_{kc0_u}^*}{2}\right) \tag{44}$$

$$v_{kq}^* = \sqrt{\frac{2}{3}}\left(\frac{\sqrt{3}}{2}v_{kb0_u}^* - \frac{\sqrt{3}}{2}v_{kc0_u}^*\right) \tag{45}$$

$$v_{o1}^* = -\sum_{n=a}^{c} v_{1n0_1}^* + \sum_{n=a}^{c} v_{2n0_1}^* + \sum_{n=a}^{c} v_{3n0_2}^* - \sum_{n=a}^{c} v_{4n0_2}^* \tag{46}$$

where $u = 1$ if $k = 1, 2$ and $u = 2$ if $k = 3, 4$.

The conduction states of the power switches, can be obtained directly from reference pole voltages ($v_{1a0_1}^*$ to $v_{4c0_2}^*$). Thus, twelve equations are necessary to calculate the PWM and consequently (44)-(46) are not sufficient to determine the pole voltages. Then, three auxiliary variables (v_x^*, v_y^* and v_z^*) are introduced:

$$v_\xi^* = \frac{1}{3}\left(v_{ka0_u}^* + v_{kb0_u}^* + v_{kc0_u}^*\right) \tag{47}$$

where $\xi = x, y, z$ with $k = 1$ if $\xi = x$, $k = 2$ if $\xi = y$ and $k = 4$ if $\xi = z$.

Writing the equations (44) to (47) in matrix form, we get

$$\mathbf{V}_{control} = \mathbf{M}\mathbf{V}_{pole} \tag{48}$$

where $\mathbf{V}_{control}$ is the reference voltages vector and $\mathbf{V}_{pole}$ is the reference pole voltages vector, with

$$\mathbf{V}_{control} = \begin{bmatrix} v_{1d}^* \\ v_{1q}^* \\ v_{2d}^* \\ v_{2q}^* \\ v_{3d}^* \\ v_{3q}^* \\ v_{4d}^* \\ v_{4q}^* \\ v_{o1}^* \\ v_x^* \\ v_y^* \\ v_z^* \end{bmatrix} \quad \text{and} \quad \mathbf{V}_{pole} = \begin{bmatrix} v_{1a0_1}^* \\ v_{1b0_1}^* \\ v_{1c0_1}^* \\ v_{2a0_1}^* \\ v_{2b0_1}^* \\ v_{2c0_1}^* \\ v_{3a0_2}^* \\ v_{3b0_2}^* \\ v_{3c0_2}^* \\ v_{4a0_2}^* \\ v_{4b0_2}^* \\ v_{4c0_2}^* \end{bmatrix}$$

and

$$\mathbf{M} = \begin{bmatrix} \sqrt{\frac{2}{3}} & -\frac{\sqrt{6}}{6} & -\frac{\sqrt{6}}{6} & 0 & 0 & 0 & 0 & 0 & 0 & 0 & 0 & 0 \\ 0 & \frac{\sqrt{2}}{2} & -\frac{\sqrt{2}}{2} & 0 & 0 & 0 & 0 & 0 & 0 & 0 & 0 & 0 \\ 0 & 0 & 0 & \sqrt{\frac{2}{3}} & -\frac{\sqrt{6}}{6} & -\frac{\sqrt{6}}{6} & 0 & 0 & 0 & 0 & 0 & 0 \\ 0 & 0 & 0 & 0 & \frac{\sqrt{2}}{2} & -\frac{\sqrt{2}}{2} & 0 & 0 & 0 & 0 & 0 & 0 \\ 0 & 0 & 0 & 0 & 0 & 0 & \sqrt{\frac{2}{3}} & -\frac{\sqrt{6}}{6} & -\frac{\sqrt{6}}{6} & 0 & 0 & 0 \\ 0 & 0 & 0 & 0 & 0 & 0 & 0 & \frac{\sqrt{2}}{2} & -\frac{\sqrt{2}}{2} & 0 & 0 & 0 \\ 0 & 0 & 0 & 0 & 0 & 0 & 0 & 0 & 0 & \sqrt{\frac{2}{3}} & -\frac{\sqrt{6}}{6} & -\frac{\sqrt{6}}{6} \\ 0 & 0 & 0 & 0 & 0 & 0 & 0 & 0 & 0 & 0 & \frac{\sqrt{2}}{2} & -\frac{\sqrt{2}}{2} \\ -1 & -1 & -1 & 1 & 1 & 1 & 1 & 1 & 1 & -1 & -1 & -1 \\ \frac{1}{3} & \frac{1}{3} & \frac{1}{3} & 0 & 0 & 0 & 0 & 0 & 0 & 0 & 0 & 0 \\ 0 & 0 & 0 & \frac{1}{3} & \frac{1}{3} & \frac{1}{3} & 0 & 0 & 0 & 0 & 0 & 0 \\ 0 & 0 & 0 & 0 & 0 & 0 & 0 & 0 & 0 & \frac{1}{3} & \frac{1}{3} & \frac{1}{3} \end{bmatrix}$$

Multiplying (48) by inverse of $\mathbf{M}$, i.e., $\mathbf{M}^{-1}$, we conclude that

$$\mathbf{V}_{pole} = \mathbf{M}^{-1}\mathbf{V}_{control}$$

where

$$M^{-1} = \begin{bmatrix}
\sqrt{\frac{2}{3}} & 0 & 0 & 0 & 0 & 0 & 0 & 0 & 0 & 1 & 0 & 0 \\
-\frac{\sqrt{6}}{6} & \frac{\sqrt{2}}{2} & 0 & 0 & 0 & 0 & 0 & 0 & 0 & 1 & 0 & 0 \\
-\frac{\sqrt{6}}{6} & -\frac{\sqrt{2}}{2} & 0 & 0 & 0 & 0 & 0 & 0 & 0 & 1 & 0 & 0 \\
0 & 0 & \sqrt{\frac{2}{3}} & 0 & 0 & 0 & 0 & 0 & 0 & 0 & 1 & 0 \\
0 & 0 & -\frac{\sqrt{6}}{6} & \frac{\sqrt{2}}{2} & 0 & 0 & 0 & 0 & 0 & 0 & 1 & 0 \\
0 & 0 & -\frac{\sqrt{6}}{6} & -\frac{\sqrt{2}}{2} & 0 & 0 & 0 & 0 & 0 & 0 & 1 & 0 \\
0 & 0 & 0 & 0 & \sqrt{\frac{2}{3}} & 0 & 0 & 0 & \frac{1}{3} & 1 & -1 & 1 \\
0 & 0 & 0 & 0 & -\frac{\sqrt{6}}{6} & \frac{\sqrt{2}}{2} & 0 & 0 & \frac{1}{3} & 1 & -1 & 1 \\
0 & 0 & 0 & 0 & -\frac{\sqrt{6}}{6} & -\frac{\sqrt{2}}{2} & 0 & 0 & \frac{1}{3} & 1 & -1 & 1 \\
0 & 0 & 0 & 0 & 0 & 0 & \sqrt{\frac{2}{3}} & 0 & 0 & 0 & 0 & 1 \\
0 & 0 & 0 & 0 & 0 & 0 & -\frac{\sqrt{6}}{6} & \frac{\sqrt{2}}{2} & 0 & 0 & 0 & 1 \\
0 & 0 & 0 & 0 & 0 & 0 & -\frac{\sqrt{6}}{6} & -\frac{\sqrt{2}}{2} & 0 & 0 & 0 & 1
\end{bmatrix}$$

Since the above equation is very important, it will be written in component form, then the reference pole voltages are represented by following set of equations:

$$v^*_{1n0_1} = v^*_{1n} + v^*_x \tag{49}$$

$$v^*_{2n0_1} = v^*_{2n} + v^*_y \tag{50}$$

$$v^*_{3n0_2} = v^*_{3n} + \frac{v^*_{o1}}{3} - v^*_y + v^*_z + v^*_x \tag{51}$$

$$v^*_{4n0_2} = v^*_{4n} + v^*_z \tag{52}$$

where v^*_{1n} to v^*_{4n} are obtained from v^*_{1dq} to v^*_{4dq} and *odq* transform with:

$$v^*_{ka} = \sqrt{\frac{2}{3}} v^*_{kd} \tag{53}$$

$$v^*_{kb} = \sqrt{\frac{2}{3}} \left(-\frac{1}{2} v^*_{kd} + \frac{\sqrt{3}}{2} v^*_{kq} \right) \tag{54}$$

$$v^*_{kc} = \sqrt{\frac{2}{3}} \left(-\frac{1}{2} v^*_{kd} - \frac{\sqrt{3}}{2} v^*_{kq} \right). \tag{55}$$

The auxiliary voltages can be independently chosen, provided that the maximum and minimum limits of the pole voltages ($v_{C_u}/2$ and $-v_{C_u}/2$) are respected, i.e.:

$$v^*_{\xi\,\max} = \frac{v^*_{C_u}}{2} - v^*_{\xi 1\,\max} \tag{56}$$

$$v^*_{\xi\,\min} = -\frac{v^*_{C_u}}{2} - v^*_{\xi 1\,\min} \tag{57}$$

where $v^*_{C_u}$ is the reference dc-link voltage, $v^*_{\xi 1\,\max} = \max\{V^*_{\xi 1}\}$, $v^*_{\xi 1\,\min} = \min\{V^*_{\xi 1}\}$, with $V^*_{x1} = \{v^*_{1a}, v^*_{1b}, v^*_{1c}, v^*_{3a} + \frac{v^*_{o1}}{3} - v^*_y + v^*_z, v^*_{3b} + \frac{v^*_{o1}}{3} - v^*_y + v^*_z, v^*_{3c} + \frac{v^*_{o1}}{3} - v^*_y + v^*_z\}$, $V^*_{y1} = \{v^*_{2a}, v^*_{2b}, v^*_{2c}\}$ and $V^*_{z1} = \{v^*_{4a}, v^*_{4b}, v^*_{4c}\}$.

Introducing a parameter μ ($0 \leq \mu \leq 1$), the variables v^*_x, v^*_y and v^*_z can be written as in [24,25]:

$$v^*_\xi = \mu_\xi v^*_{\xi\,\max} + (1 - \mu_\xi) v^*_{\xi\,\min} \tag{58}$$

where $0 \leq \mu_\xi \leq 1$.

Note that, when the maximum ($\mu = 1$) or minimum ($\mu = 0$) value are selected, one of the converter legs operates with zero switching frequency [see Figs. 3(c) and (d)]. On the other hand, operation with the average values ($\mu = 0.5$) generates pulse voltage centered in the sampling period, has in general an improving in the harmonic distortion of the voltages. The pole voltage with $\mu = 0.5$ are shown in Figs. 3(a) and (b).

Once v^*_x, v^*_y and v^*_z are chosen, twelve pole voltages ($v^*_{1n0_1}$, $v^*_{2n0_1}$, $v^*_{3n0_2}$ and $v^*_{4n0_2}$) are defined from (49)-(52). The gating signals can be obtained comparing pole voltages with a high frequency triangular carrier signal. The gating signals were obtained comparing pole voltages with a single or double high frequency triangular carrier signals, i.e., single or double-carrier-based PWM. With double-carrier based PWM, the phase shift between the triangular carrier signals is 180°. In this case the reference pole voltages of the converters 1 and 2 are compared with a carrier PWM and the reference pole voltages of the converters 3 and 4 are compared with the other carrier PWM shift of 180°.

The auxiliary voltages change the distribution of the null-vectors within the same sampling interval like as non-sinusoidal modulating signals (NSMS). This is beneficial since it is already known that the use of certain NSMS permits to increase the switching frequency of the inverter and also the total harmonic distortion can be decreased [24, 25]. Different types of NSMS are obtained depending on how the null-vectors are distributed within the sampling interval, that is, as a function of the parameter μ defined by auxiliary variables.

Fig. 3 shows four possibility of some of the well-known distorted modulating signals for three different values of μ, with single and double carrier PWM. Fig. 3 shows the waveforms of: reference pole voltages ($v^*_{1a0_1}$ and $v^*_{3a0_2}$), carriers PWM (v_{t1} and v_{t2}), instantaneous pole voltages (v_{1a0_1} and v_{3a0_2}), average voltages generated by converters 1 and 3 [$v_a = (v_{1a0_1} - v_{g0_1} + v_{3a0_2} - v_{g0_2})/2$] and circulating voltage (v_{o1}).

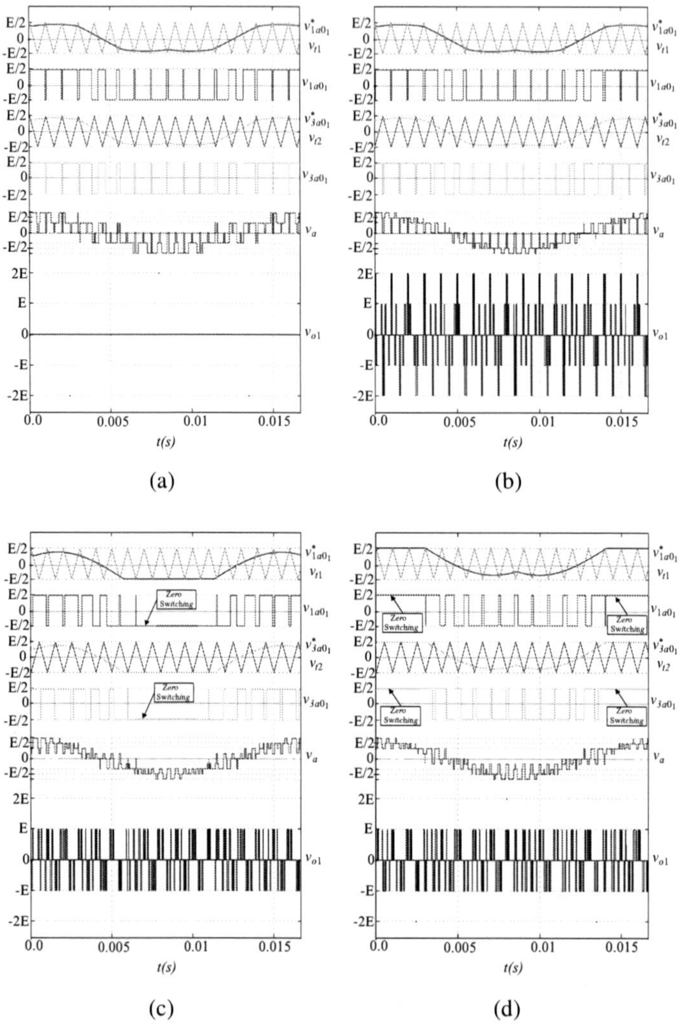

Figure 3. Analysis of PWM signals for Topology T2B. (a) Single-Carrier $\mu = 0.5$. (b) Double-Carrier $\mu = 0.5$. (c) Double-Carrier $\mu = 0$. (d) Double-Carrier $\mu = 1$.

Fig. 3(a) shows results with single-carrier and $\mu = 0.5$. When $\mu = 0.5$, the symmetrical PWM and the null-vectors are applied at the beginning and end of the sampling period. Fig. 3(b) shows the same results but with double-carrier. Due to the interleaved technique the converter switching pattern is changed for the converter 3, such benefit is the increase in the number of levels of the average generated voltage (v_a) decrease the harmonic distortion of the grid currents. However, in this case there is circulating voltage, which increases the harmonic distortion of internal currents.

The Figs. 3(c) and (d) show the waveforms with $\mu = 0$ and $\mu = 1$ using double-carrier PWM, respectively. It is notable that the reference pole voltages are equal to discontinuous PWM (DPWM) [24, 26]. When $\mu = 0$ or $\mu = 1$ the null-vectors are distributed at the beginning or end of the sampling period, respectively. The leading feature of DPWM is that one of the phases is clamped by $1/3$ of the period of the fundamental and the number of commutations per cycle of the power switches is reduced to identical carrier frequency. Fewer commutations per cycle of the power switches leads to less switching losses. In addition, with double-carrier PWM has increase in the number of levels of the average generated voltage (v_a) and a better distribution of vectors. Another advantage is the reduction of the circulating voltage. Less circulating voltage leads the reduction of harmonic distortion of internal currents and decrease of the ripple of the circulating currents. The analysis and conclusions reached are similar for converters 2 and 4.

3.2. Configuration T1B

The reference pole voltages should be calculated from the reference voltages v^*_{kd}, v^*_{kq}, v^*_{o1} and v^*_{o2} defined by the input and output controllers (see Section 4), that is

$$v^*_{kd} = \sqrt{\frac{2}{3}}\left(v^*_{ka0} - \frac{v^*_{kb0}}{2} - \frac{v^*_{kc0}}{2}\right) \tag{59}$$

$$v^*_{kq} = \sqrt{\frac{2}{3}}\left(\frac{\sqrt{3}}{2}v^*_{kb0} - \frac{\sqrt{3}}{2}v^*_{kc0}\right) \tag{60}$$

$$v^*_{o1} = -v^*_{1a0} - v^*_{1b0} - v^*_{1c0} + v^*_{3a0} + v^*_{3b0} + v^*_{3c0} \tag{61}$$

$$v^*_{o2} = v^*_{2a0} + v^*_{2b0} + v^*_{2c0} - v^*_{4a0} - v^*_{4b0} - v^*_{4c0}. \tag{62}$$

There are ten reference voltages, defined by (59) to (62). Such equations are not sufficient to determine the twelve reference pole voltages v^*_{1n0} to v^*_{4n0}, then two auxiliary variables v^*_x and v^*_y are introduced defined by (47) with $k = 1$ if $\xi = x$ and $k = 2$ if $\xi = y$.

Writing the equations (59) to (62) and equation (47) in matrix form, we get

$$\mathbf{V}_{control} = \mathbf{M}\mathbf{V}_{pole} \tag{63}$$

where $\mathbf{V}_{control}$ is the reference voltages vector and $\mathbf{V}_{pole}$ is the reference pole voltages vector, with

$$\mathbf{V}_{control} = \begin{bmatrix} v_{1d}^* \\ v_{1q}^* \\ v_{2d}^* \\ v_{2q}^* \\ v_{3d}^* \\ v_{3q}^* \\ v_{4d}^* \\ v_{4q}^* \\ v_{o1}^* \\ v_{o2}^* \\ v_x^* \\ v_y^* \end{bmatrix} \quad \text{and} \quad \mathbf{V}_{pole} = \begin{bmatrix} v_{1a0}^* \\ v_{1b0}^* \\ v_{1c0}^* \\ v_{2a0}^* \\ v_{2b0}^* \\ v_{2c0}^* \\ v_{3a0}^* \\ v_{3b0}^* \\ v_{3c0}^* \\ v_{4a0}^* \\ v_{4b0}^* \\ v_{4c0}^* \end{bmatrix}$$

and

$$\mathbf{M} = \begin{bmatrix} \sqrt{\frac{2}{3}} & -\frac{\sqrt{6}}{6} & -\frac{\sqrt{6}}{6} & 0 & 0 & 0 & 0 & 0 & 0 & 0 & 0 & 0 \\ 0 & \frac{\sqrt{2}}{2} & -\frac{\sqrt{2}}{2} & 0 & 0 & 0 & 0 & 0 & 0 & 0 & 0 & 0 \\ 0 & 0 & 0 & \sqrt{\frac{2}{3}} & -\frac{\sqrt{6}}{6} & -\frac{\sqrt{6}}{6} & 0 & 0 & 0 & 0 & 0 & 0 \\ 0 & 0 & 0 & 0 & \frac{\sqrt{2}}{2} & -\frac{\sqrt{2}}{2} & 0 & 0 & 0 & 0 & 0 & 0 \\ 0 & 0 & 0 & 0 & 0 & 0 & \sqrt{\frac{2}{3}} & -\frac{\sqrt{6}}{6} & -\frac{\sqrt{6}}{6} & 0 & 0 & 0 \\ 0 & 0 & 0 & 0 & 0 & 0 & 0 & \frac{\sqrt{2}}{2} & -\frac{\sqrt{2}}{2} & 0 & 0 & 0 \\ 0 & 0 & 0 & 0 & 0 & 0 & 0 & 0 & 0 & \sqrt{\frac{2}{3}} & -\frac{\sqrt{6}}{6} & -\frac{\sqrt{6}}{6} \\ 0 & 0 & 0 & 0 & 0 & 0 & 0 & 0 & 0 & 0 & \frac{\sqrt{2}}{2} & -\frac{\sqrt{2}}{2} \\ -1 & -1 & -1 & 0 & 0 & 0 & 1 & 1 & 1 & 0 & 0 & 0 \\ 0 & 0 & 0 & 1 & 1 & 1 & 0 & 0 & 0 & -1 & -1 & -1 \\ \frac{1}{3} & \frac{1}{3} & \frac{1}{3} & 0 & 0 & 0 & 0 & 0 & 0 & 0 & 0 & 0 \\ 0 & 0 & 0 & \frac{1}{3} & \frac{1}{3} & \frac{1}{3} & 0 & 0 & 0 & 0 & 0 & 0 \end{bmatrix}$$

Multiplying (63) by inverse of $\mathbf{M}$, i.e., $\mathbf{M}^{-1}$, we conclude that

$$\mathbf{V}_{pole} = \mathbf{M}^{-1}\mathbf{V}_{control}$$

where

$$M^{-1} = \begin{bmatrix}
\sqrt{\frac{2}{3}} & 0 & 0 & 0 & 0 & 0 & 0 & 0 & 0 & 0 & 1 & 0 \\
-\frac{\sqrt{6}}{6} & \frac{\sqrt{2}}{2} & 0 & 0 & 0 & 0 & 0 & 0 & 0 & 0 & 1 & 0 \\
-\frac{\sqrt{6}}{6} & -\frac{\sqrt{2}}{2} & 0 & 0 & 0 & 0 & 0 & 0 & 0 & 0 & 1 & 0 \\
0 & 0 & \sqrt{\frac{2}{3}} & 0 & 0 & 0 & 0 & 0 & 0 & 0 & 0 & 1 \\
0 & 0 & -\frac{\sqrt{6}}{6} & \frac{\sqrt{2}}{2} & 0 & 0 & 0 & 0 & 0 & 0 & 0 & 1 \\
0 & 0 & -\frac{\sqrt{6}}{6} & -\frac{\sqrt{2}}{2} & 0 & 0 & 0 & 0 & 0 & 0 & 0 & 1 \\
0 & 0 & 0 & 0 & \sqrt{\frac{2}{3}} & 0 & 0 & 0 & \frac{1}{3} & 0 & 1 & 0 \\
0 & 0 & 0 & 0 & -\frac{\sqrt{6}}{6} & \frac{\sqrt{2}}{2} & 0 & 0 & \frac{1}{3} & 0 & 1 & 0 \\
0 & 0 & 0 & 0 & -\frac{\sqrt{6}}{6} & -\frac{\sqrt{2}}{2} & 0 & 0 & \frac{1}{3} & 0 & 1 & 0 \\
0 & 0 & 0 & 0 & 0 & 0 & \sqrt{\frac{2}{3}} & 0 & 0 & -\frac{1}{3} & 0 & 1 \\
0 & 0 & 0 & 0 & 0 & 0 & -\frac{\sqrt{6}}{6} & \frac{\sqrt{2}}{2} & 0 & -\frac{1}{3} & 0 & 1 \\
0 & 0 & 0 & 0 & 0 & 0 & -\frac{\sqrt{6}}{6} & -\frac{\sqrt{2}}{2} & 0 & -\frac{1}{3} & 0 & 1
\end{bmatrix}$$

Since that above equation is very important, we writing in component form, then the reference pole voltages are represented by following set of equations:

$$v_{1n0}^* = v_{1n}^* + v_x^* \quad (64)$$

$$v_{2n0}^* = v_{2n}^* + v_y^* \quad (65)$$

$$v_{3n0}^* = v_{3n}^* + \frac{v_{o1}^*}{3} + v_x^* \quad (66)$$

$$v_{4n0}^* = v_{4n}^* - \frac{v_{o2}^*}{3} + v_y^* \quad (67)$$

with the reference voltages v_{1n}^* to v_{4n}^* obtained by (53) to (55).

The variables v_x^* and v_y^* are defined from (56)-(57), with the voltages $V_{x1}^* = \{v_{1a}^*, v_{1b}^*, v_{1c}^*, v_{3a}^* + v_{o1}^*/3, v_{3b}^* + v_{o1}^*/3, v_{3c}^* + v_{o1}^*/3\}$ and $V_{y1}^* = \{v_{2a}^*, v_{2b}^*, v_{2c}^*, v_{4a}^* - v_{o2}^*/3, v_{4b}^* - v_{o2}^*/3, v_{4c}^* - v_{o2}^*/3\}$. Similarly the auxiliary variables can be whiten as a function of μ from (58).

4. Control Strategy

There are four main control goals on the proposed configurations, they are: regulated load voltage, power factor control with sinusoidal grid current, dc-link voltage regulation and cancelation of the circulating current. Figure 4(a) shows as the control block diagram achieve those objectives for the configuration T2B. Since this topology employs two dc-links, the voltages v_{c1} and v_{c2} are regulated by the controllers R_{c1} and

R_{c2}. Those controllers furnish the amplitudes of the reference currents I_1^* and I_3^* required at the input of the converters 1 and 3. In turn, the power factor control is obtained by synchronization of the reference grid currents i_{1abc}^* (i_{1a}^*, i_{1b}^*, i_{1c}^*) and i_{3abc}^* (i_{3a}^*, i_{3b}^*, i_{3c}^*) with the three-phase voltages. Such a synchronization is obtained with G_{ir}, which employs a PLL based technique.

The output of the block G_{ir} are instantaneous currents that are converted to dq variables through the transformation *123/dq*, based on the block *dq* [see equation (72)]. As a consequence the output of these blocks are the reference currents i_{1dq}^* (i_{1d}^*, i_{1q}^*) and i_{3dq}^* (i_{3d}^* and i_{3q}^*). A double-sequence controller type is employed on R_{i1dq} and R_{i3dq} to regulate the sinusoidal currents i_{1dq}^* and i_{3dq}^*. The output of the controllers R_{i1dq} and R_{i3dq} will define the reference voltages v_{1dq}^* and v_{3dq}^*.

The circulating current is regulated by R_{o1} with $i_{o1}^* = 0$. The output of the block R_{o1} is circulating voltage v_{o1}^*.

The load voltage control is also established by a double-sequence type of controller via block R_{vdq}. The output of this controller defines v_{dq}^* (v_d^* and v_q^*). The voltage applied for each converter can be considered equal to the voltage v_{dq}, which turns out that the output of the block G_{v24} is $v_{2dq}^* = v_{4qd}^* = v_{dq}^*$.

The reference voltages generated by the controllers v_{1dq}^*, v_{3dq}^*, v_{o1}^*, v_{2dq}^* and v_{4dq}^*, and the auxiliary voltages v_x^*, v_y^* and v_z^* [defined by (58)], are applied to the block PWM. The output of this block generates the state of the switches.

The main differences between the control strategies presented earlier for the configuration T2B and that one for the topology T1B [see Fig. 4(b)] are: *i)* just a single controller is employed to regulate the dc-link voltage with its output defining the amplitude of the grid currents; *ii)* the amplitudes of the currents i_{g1} and i_{g2} are both equal to $I_g/2$; *iii)* it is necessary control two circulating currents instead of just one as in configuration T2B.

5. WTHD Analysis

The Fig. 5 shows the WTHD for the voltages v_{gd} and v_{gq} of the topologies T2B, T1B and conventional one. The WTHD are plotted as a function of the μ introduced in the PWM section. It is worth mentioning that the of the WTHD for the voltages v_{ld} and v_{lq} have similar characteristics and so will be omitted.

From this figure it is possible to state that the PWM of the voltages v_{gd} and v_{gq}, by using one carrier signal is equal to the WTHD of the conventional configuration. On the other hand, the WTHD of the PWM with carrier signals is always for the proposed configurations regardless the value of μ. The reduction of the WTHD (with $\mu = 0.5$) is almost 60% as compared to the conventional topology with its lowest value of WTHD (i.e., $\mu = 0.5$).

An explanation for the behavior of the voltages v_{gd} and v_{gq} can be obtained from the Fig. 6. In this figure the v_{gd} is presented in the time domain for both conventional

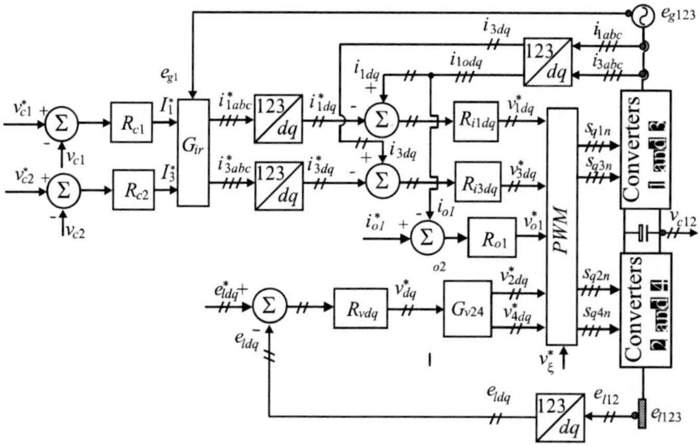

(a)

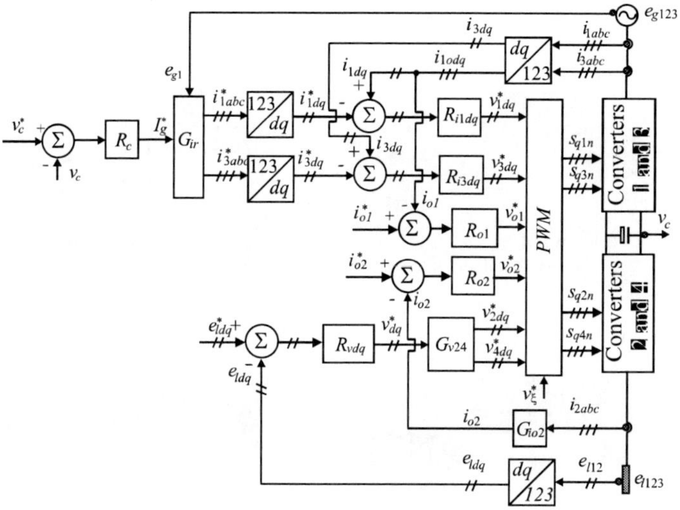

(b)

Figure 4. Control block diagram. (a) Configuration T2B. (b) Configuration T1B.

topology and T2B with $\mu = 0.5$. Notice that the voltage v_{gd}, by using only one carrier signal with $\mu = 0.5$, is equal to the WTHD of the v_{gd} for the conventional topology. On the other hand, with two carrier signals the number of levels of the voltage v_{gd} increases

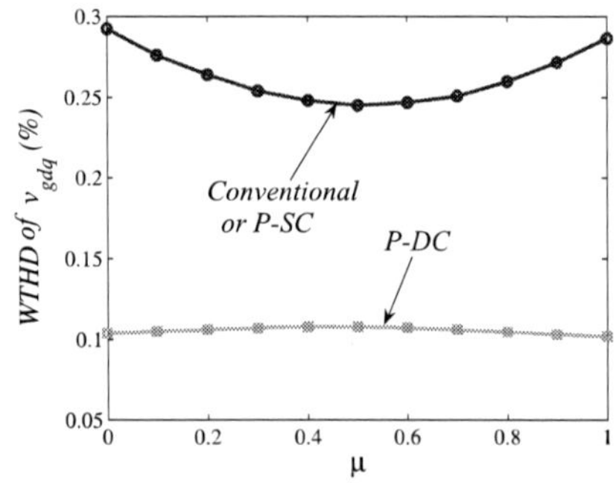

Figure 5. Average value of WTHD for the voltages v_{gd} and v_{gq} as a function of μ - Configuration T2B.

as compared to the conventional configuration. As a consequence the WTHD will be improved.

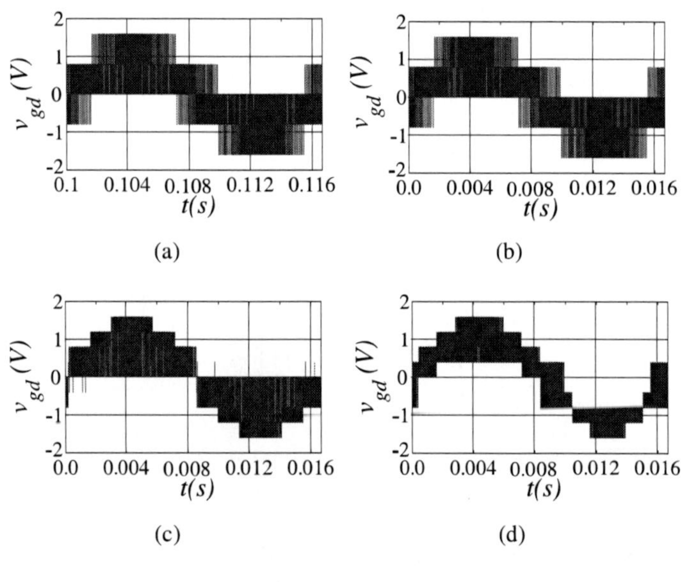

Figure 6. Voltage v_{gd} .(a) Conventional configuration with $\mu = 0.5$. (b) Topology T2B by using only one carrier signal with $\mu = 0.5$. (c) Topology T2B by using two carrier signals with $\mu = 0.5$. (d) Topology T2B by using two carrier signals with $\mu = 0$.

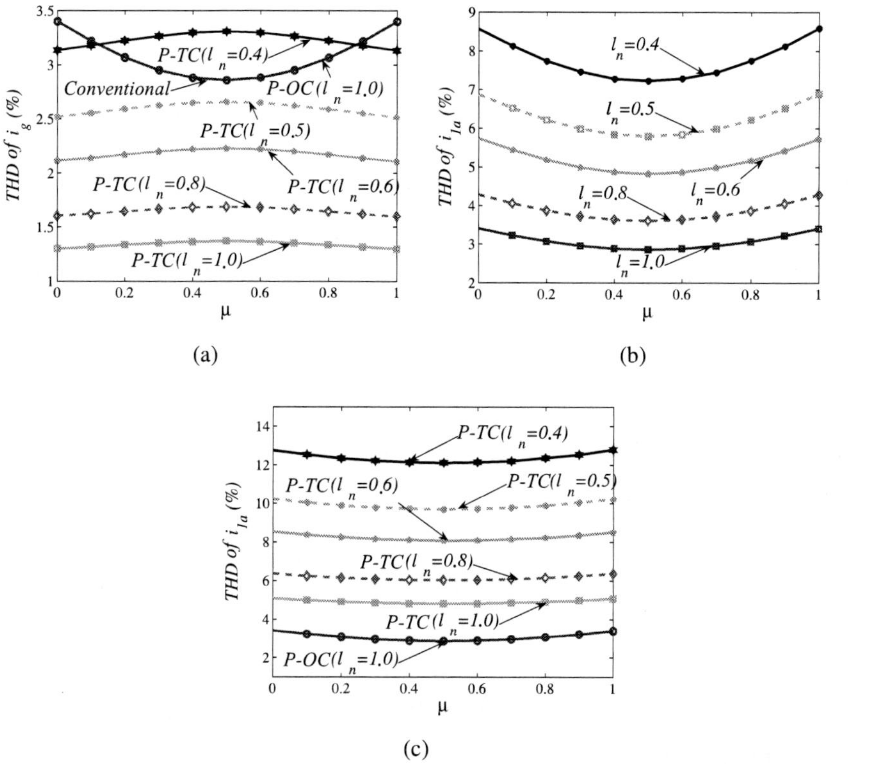

Figure 7. THD as a function of μ for different values of l_n: (a) Grid current for phase 1 (i_{g1}), (b) internal current i_{1a} (Configuration T2B), (c) internal current i_{1a} (Configuration T1B).

6. Harmonic Distortion Analysis

The harmonic distortion for the grid and internal currents will be presented in this section for different values of filter inductors. Notice that the WHTD (Weighted Harmonic Total Distortion) for the voltages v_{gd} and v_{gq} will define the THD of the grid current.

The Fig. 7 illustrates the THD of the: grid current for phase 1 (i_{g1}) and internal current (i_{1a}) as a function of μ for different values of l_n. l_n is a relationship of the equivalent inductance of topologies T2B and T1B and the inductance of the conventional configuration, i.e., $l_n = l_g(P)/l_g(C)$]. The same behavior is observed for the other two phases.

The THD of the current i_{g1} with only one carrier PWM signal and $l_n = 1.0$ are equivalent the THD of the current i_{g1} for the conventional topology, as observed in Fig. 7(a). In this case $l_g(P) = l_g(C)$, which means that the inductance of the configuration T2B is twice as compared to the conventional one ($l_1 = 2l_g(C)$). On the other hand, the PWM with two carrier signals allows a reduction on the THD of the current i_{g1} for the topology T2B when $l_n \geq 0.5$, which means $l_1 = l_g(C)$ (see Fig. 7(a)). The THD of the current i_{1a} for the configuration T2B with one carrier signal are equivalent to those ones with two carrier signals when $l_n = 1.0$. Also, the THD of the internal currents (i_{1a}) are similar to the THD of the current i_{g1} for the conventional configuration. However, the THD of the internal current (i_{1a}) experiences an increase when $l_n < 1.0$, as presented in Fig. 7(b).

The THD of the internal current (i_{1a}) of the topology T1B is shown in Fig. 7(c). Notice that for one carrier signal, the THD of the current i_{1a} is equal to the THD of the grid current. However, the THD with two carrier signals is higher. When $l_n < 1$, there is an increase on the THD of the internal currents, as shown in Fig. 7(c).

For the topology T1B with $\mu = 0$, the THD of the current i_{g1} is lower than the conventional one when $l_n = 0.5$, the THD of the current i_{g1} of the topology T1B, with $\mu = 0$, is lower than the THD of the current i_{g1} for the conventional topology with $\mu = 0.5$ (it is observed a reduction of almost 13% on THD value). However, in this case the THD of the current i_{1a} is almost 4.5 higher than the THD for the conventional topology (i_{g1}).

7. THD Analysis as a Function of the Switching Frequency

Fig. 8 shows the THD of the current i_{g1} by using two carrier signals as a function of μ and considering different values of switching frequency.

From these results it is clear that a PWM with two carrier signals allows a THD reduction of i_{g1} for the proposed configurations when the switching frequency is higher than 5 kHz.

8. Semiconductor Losses

The losses estimation was obtained by using an approach presented by [27]. The power switch models were obtained experimentally based on both instantaneous losses and regression analysis. Such a regression model was employed to adjust the discrete points obtained for the switching and conduction losses. The power switch used for the semiconductor losses study was CM50DY-24H (POWEREX), with the SEMIKRON driver SKHI-10.

Fig. 9 shows the estimated losses for the configurations T2B and T1B. Those results were obtained for three different power conditions, i.e.: rated power, 75% of the rated

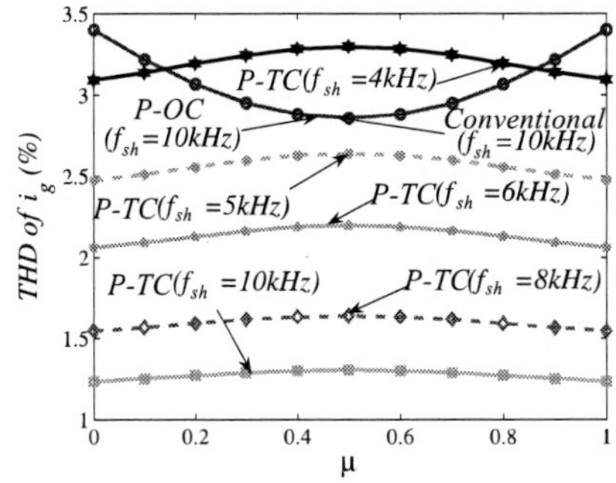

Figure 8. THD of the current i_{g1} for different values of the switching frequency as a function of μ.

power and 50% of the rated power. The losses presented in these figures are: conduction losses (P_{cd}), switching losses (P_{sw}) and total losses ($P_t = P_{cd} + P_{sw}$).

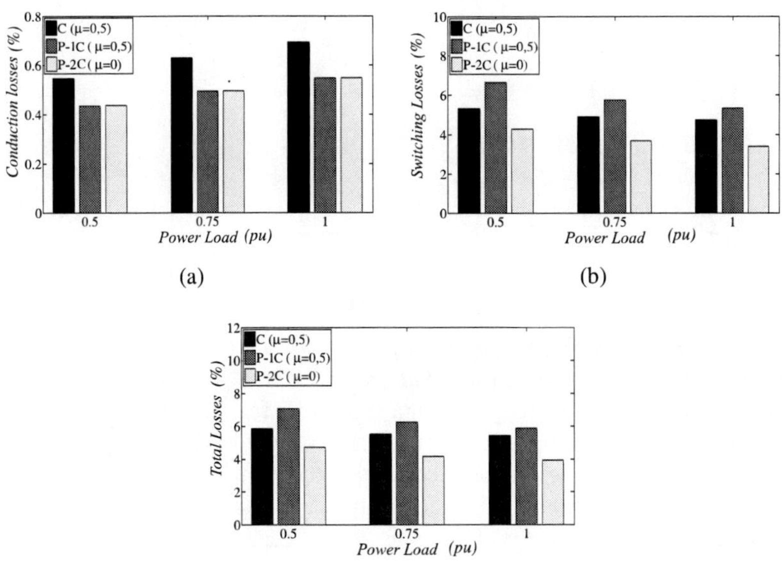

Figure 9. Estimated losses for the converters T2B and T1B. (a) Conduction losses P_{cd}. (b) Switching losses P_{sw}. (c) Total losses $P_t = P_{sw} + P_{ch}$.

It is worth mentioning that the conduction losses are always lower for the proposed topologies [Fig. 9(a)]. On the other hand, the switching losses with one carrier-signal PWM with $\mu = 0.5$ are larger than the conventional configuration. However, the PWM strategy with two carrier-signals and $\mu = 0$ reduces the switching losses [see Fig. 9(b)], which represents the biggest reduction in terms of losses.

The Table 1 shows the total losses for configurations T2B and T1B, normalized as a function of the total losses of the conventional topology $[P_t(P)/P_t(C)]$. Notice that with two-carrier signals PWM and with $\mu = 0$, the total losses reduction for the rated power is about 36%.

Table 1. Total losses for the configurations T2B and T1B normalized with respect to the conventional topology

Total losses $[P_t(P)/P_t(C)]$				
PWM Strategies		**Rated power**		
		50%	75%	100%
$P-1C$	$(\mu = 0.5)$	1.21	1.13	1.08
$P-2C$	$(\mu = 0)$	0.81	0.76	0.73

9. Dc Link Capacitor Variables

The minimum capacitance value needed for a dc-link capacitor bank can be defined as below [4, 17]:

$$C_j = \frac{i_{cj}\Delta t}{\Delta v_{cj}} \tag{68}$$

with $j = 1, 2$, $i_{cj} = s_{q\delta a} i_{\delta a} + s_{q\delta b} i_{\delta b} - s_{q\epsilon a} i_{\epsilon a} - s_{q\epsilon b} i_{\epsilon b}$ and Δv_{cj} are, the current and the dc-link voltage oscillation, respectively, with $\delta = 1, 3$ and $\epsilon = \delta + 1$.

The amount of energy stored into the capacitors for the configuration T2B is the same as the conventional topology, which leads for half of the energy in each capacitance bank. As the voltages for both converters (conventional and T2B) are the same ($v_{c1} = v_{c2} = v_c$), the currents of the conventional configuration is half of the conventional circuit ($i_{c1} = i_{c2} = i_c/2$). This leads for two capacitances half of that employed in the conventional configuration ($C_1 = C_2 = C/2$). Since one of the goals of the proposed configurations is to increase their fault tolerance capability, the capacitance employed for both dc-link will be the same as the conventional topology.

On the other hand, the total amount of energy stored in the capacitor bank for the configuration T1B is the same as in the conventional configuration, which leads for the same value of capacitance.

9.1. High Frequency Losses

Another important characteristic for dc-link capacitor design is the *RMS* capacitor current, which is responsible for the capacitor losses. The lifespan of the capacitors are reduced with the increase of this current as well as temperature. The capacitor power losses can be calculated as [28]

$$P_{c,loss} = \sum_{h=1}^{N_h} ESR(h) I_c^2(h) \tag{69}$$

where *ESR* is the *Equivalent Series Resistance* and $I_c(h)$ is the harmonic component of the capacitor. Different works in the technical literature deal with estimation of the *ESR* for a electrolytic capacitor [28–33]. The equivalent series resistance reduces with the following parameters: frequency, capacitor current high frequency oscillation, capacitor size, capacitance and environment temperature. However, the *ESR* can be considered constant for higher frequencies than 3kHz [4], [33], [34]. In fact, *ESR* can be considered equal to 45% of the equivalent series resistance measured at 100Hz [4, 34]. In this way, from equation (69), it turns out that [35]

$$P_{c,loss}^{hf} = 0.45 ESR_{(100Hz)} (I_{c,rms}^{hf})^2 \tag{70}$$

with

$$I_{c,rms}^{hf} = \sqrt{\sum_{h>50}^{N_h} I_c^2(h)} \tag{71}$$

where $P_{c,loss}^{hf}$ represents the power losses of the high frequency harmonics and $I_{c,rms}^{hf}$ is the *RMS* capacitor current for high frequency harmonics (with $h > 50$).

Since *ESR* can be considered almost constant for higher frequencies, the high frequencies losses $P_{c,loss}^{hf}$ is a function of the high frequency *RMS* capacitor current $I_{c,rms}^{hf}$. This means that the reduction in the power losses for the capacitor bank is determined by the reduction of the *RMS* capacitor current.

The Fig. 10 shows the spectrum of the capacitor current harmonics for the conventional topology i_{c1}, topology T2B and topology T1B.

Since the topology T2B has two dc-link capacitors, the power flow through each capacitor bank is half of the total power flowing for the conventional topology. Therefore, the amplitudes of the harmonic spectrums for the currents of the topology T2B (i_{c1} and i_{c2}) are equivalent to half of the that in the conventional configuration (i_c).

On the other hand, the amplitude of the spectrum for the capacitor current of the topology T1B, with a single-carrier PWM signal, are higher than the spectrum amplitudes for the conventional configuration. However, by using two carrier signals, the reduction on the spectrums are considerable higher for the proposed topologies.

Tables 2 and 3 illustrate the values of the current $I_{c,rms}^{hf}$ of the topologies T2B and T1B, normalized by the current $I_{c,rms}^{hf}$ of the conventional configuration, with $\mu =$

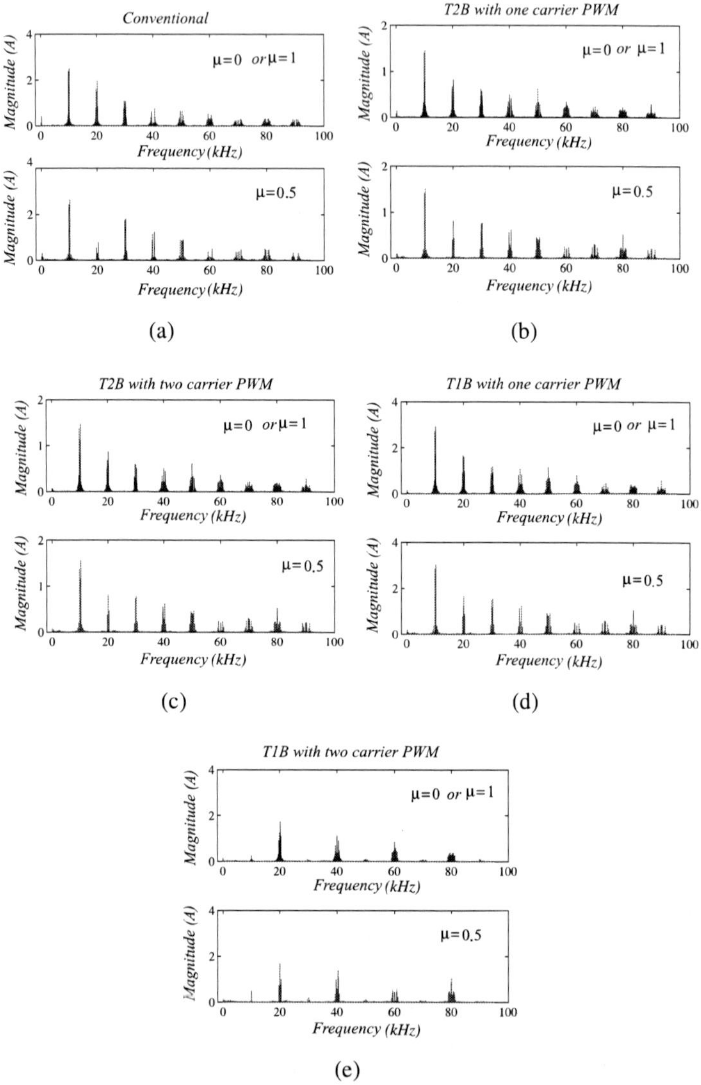

Figure 10. Spectrum of the capacitor current harmonics i_c and i_{c1}. (a) Conventional configuration. (b) Topology T2B with a single-carrier PWM signal. (c) Topology T2B with two-carrier PWM signal. (d) Topology T1B with a single-carrier PWM signal. (e) Topology T1B with two carrier PWM signals.

0.5. The currents $I^{hf}_{c,rms}$ were calculated by the equation (71) with $h > 50$, from the spectrum of the capacitor current shown in Fig. 10. For the topology T2B, the biggest reduction for the current $I^{hf}_{c,rms}$ was obtained by using one or two carrier signals, with $\mu = 0.5$. In this case, the reduction of the high frequency RMS current was close to 43%. On the other hand, for the topology T1B, the biggest reduction for $I^{hf}_{c,rms}$ was obtained with two-carrier PWM signals, with $\mu = 0.5$, the reduction of the current, in this case, was close to 38% as compared to the conventional configuration, with $\mu = 0.5$.

Table 2. Current $I^{hf}_{c,rms}$ of the topology T2B, normalized by current $I^{hf}_{c,rms}$ of the conventional configuration

Current $I^{hf}_{c,rms}(P)/I^{hf}_{c,rms}(C)$			
PWM strategy	$\mu = 0$	$\mu = 0.5$	$\mu = 1$
$P-1C$	0.61	0.57	0.61
$P-2C$	0.62	0.57	0.62

The high frequency power losses ($P^{hf}_{c,loss}$) are defined by the equation (70). Considering that the capacitor values of the topology T2B are equal to each other, then, $ESR_{100Hz}(P) = ESR_{100Hz}(C)$. Hence, with one or two carrier signals, with $\mu = 0.5$, the high frequency power losses reduction for two dc-link capacitors is about 68% as compared to the conventional topology. For the topology T1B ($ESR_{100Hz}(P) = ESR_{100Hz}(C)$), the high frequency power losses reduction with two carrier signals, with $\mu = 0.5$, was of about 61%.

Table 3. Current $I^{hf}_{c,rms}$ of the topology T1B, normalized by current $I^{hf}_{c,rms}$ of the conventional configuration

Current $I^{hf}_{c,rms}(P)/I^{hf}_{c,rms}(C)$			
PWM strategy	$\mu = 0$	$\mu = 0.5$	$\mu = 1$
$P-1C$	1.22	1.14	1.22
$P-2C$	0.72	0.62	0.72

10. Simulation Results

10.1. Topology T2B

Figs. 11 and 12 show simulation results for configuration T2B [see Fig. 1(a)], obtained for a balanced three-phase system.

The Fig. 11 illustrates the results by using a single carrier PWM signal, with $\mu = 0.5$. Notice from those results that i_{g1} is a sinusoidal waveform and it is in phase with the grid voltage of the phase 1 [see Fig. 11(a)], the circulating current is null [Fig. 11(d)], and both dc-link voltages and load voltages are under control, as presented in Figs. 11(e), 11(f) and 11(g). Due to the connection of the back-to-back converters in parallel, the converter currents are equivalents to the half of the grid current, see Figs.11(b) and 11(c).

The simulation results with two carrier-signal PWM and $\mu = 0$, are shown in Fig. 12. In this case, it is observed a better reduction of the high-frequency components of the grid currents [see Fig. 12(a) for the current i_{g1}] and of the three-phase load voltage [Fig. 12(g)]. On the other hand, $\mu = 0$ means an increase of the high-frequency components for the converter currents as well as for the circulating current [Figs. 12(b), 12(c) and 12(d)], as compared to a PWM with a single-carrier signal. As expected, the dc-link voltages are also regulated, as presented in Figs. 12(d) and 12(e)].

10.2. Topology T1B

The simulation results for the configuration T1B [Fig. 1(a)] are presented in Figs. 13 and 14.

The results presented in Fig. 13 are for a single-carrier PWM with $\mu = 0.5$. It is observed from those results that the control system operated as expected, with a sinusoidal current i_{g1} in phase with the voltage e_{g1} [see Fig. 13(a)], both circulating currents close to zero [Figs. 13(d) and 13(e)], and dc-link voltages and load voltage under control [see Figs. 13(f) and 13(h)]. Due to the parallel connection of the back-to-back converters, the amplitude of the converter currents are half of the grid currents, as in Figs. 13(b) and 13(c).

The Fig. 12 shows the simulation results by using two carrier signals with $\mu = 0$. In this case, it is observed that the grid currents and the load voltages present high-frequency components, as seen in Figs. 14(a) and 14(h). On the other hand, with $\mu = 0$, there is an increase of the high frequency components for both internal and circulating currents [Figs. 14(b), 14(c), 14(e) and 14(f)], as compared to the last case by using a single carrier signal.

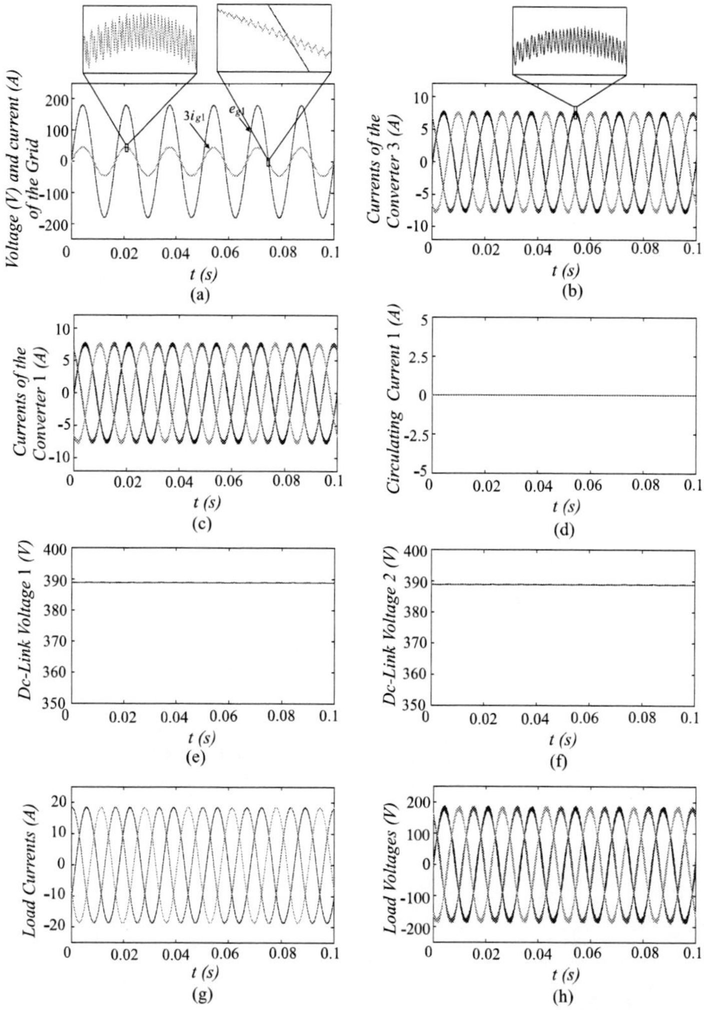

Figure 11. Simulation results for topology T2B with a single carrier PWM signal, with $\mu = 0.5$. (a) Grid voltage and current for phase 1 (e_{g1} and i_{g1}). (b) Currents of the converter 3 (i_{3a}, i_{3b} and i_{3c}). (c) Currents of the converter 1 (i_{1a}, i_{1b} and i_{1c}). (d) Circulation current (i_{o1}). (e) Dc-link voltage 1 (v_{c1}). (f) Dc-link voltage 2 (v_{c2}). (g) Load currents (i_{l1}, i_{l2} and i_{l3}). (h) Load voltages (e_{l1}, e_{l2} and e_{l3}).

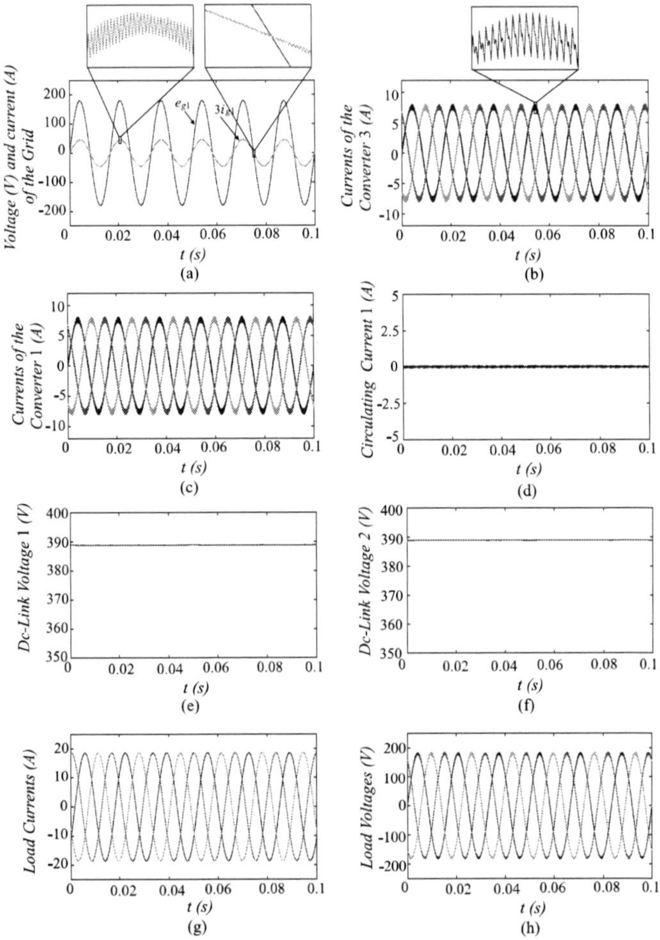

Figure 12. Simulation results for configuration T2B by using two-carrier PWM signals, with $\mu = 0$. (a) Grid voltage and current of the phase 1 e_{g1} and i_{g1}. (b) currents of the converter 3 (i_{3a}, i_{3b} and i_{3c}). (c) Currents of the converter 1 (i_{1a}, i_{1b} and i_{1c}). (d) Circulating current i_{o1}. (e) Dc-link voltage 1 (v_{c1}). (f) Dc-link voltage 2 (v_{c2}). (g) Load currents (i_{l1}, i_{l2} and i_{l3}). (h) Load voltages (e_{l1}, e_{l2} and e_{l3}).

10.3. Experimental Results

Figs. 15 and 16 show the experimental results obtained for the configuration T2B. Those results were obtained with a volts/hertz strategy for the three-phase motor

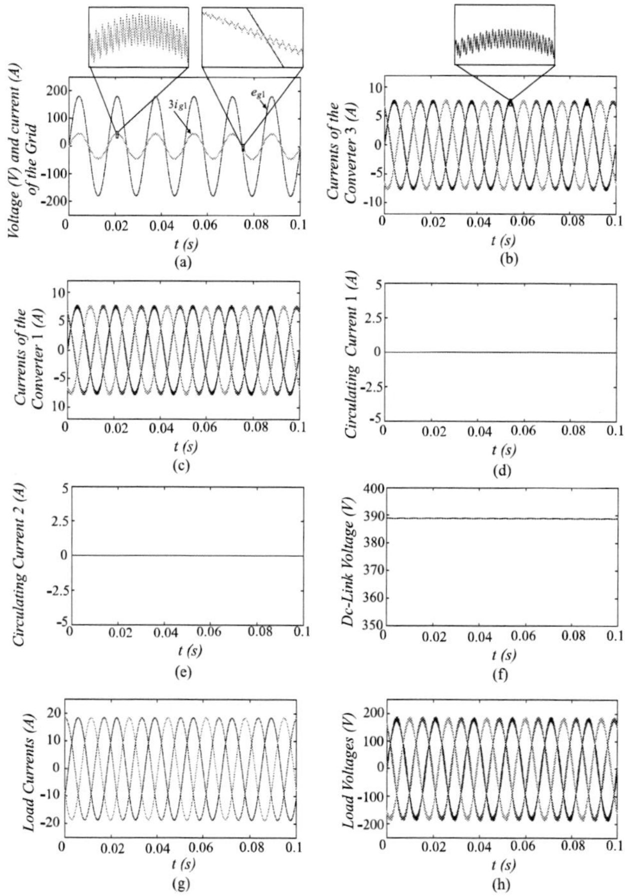

Figure 13. Simulation results for the topology T1B by using a single-carrier signal and $\mu = 0.5$. (a) Grid voltage and current for phase 1 (e_{g1} and i_{g1}). (b) Currents of the converter 1 (i_{1a}, i_{1b} and i_{1c}). (c) Currents of the converter 3 (i_{3a}, i_{3b} and i_{3c}). (d) Circulating current i_{o1}. (e) Circulating current i_{o2}. (f) Dc-link voltage (v_c). (g) Load currents (i_{l1}, i_{l2} and i_{l3}). (h) Load voltages (e_{l1}, e_{l1} and e_{l3}).

(V/Hz=100V/50Hz). The waveforms are presented in the following sequence: grid voltage and current for phase 1 (e_{g1} and i_{g1}), internal currents of converter 1 (i_{1a}, i_{1b} and i_{1c}), circulating current (i_{o1}), dc-link voltages (v_{c1} and v_{c2}) and load voltages (e_{l1} and e_{l2}).

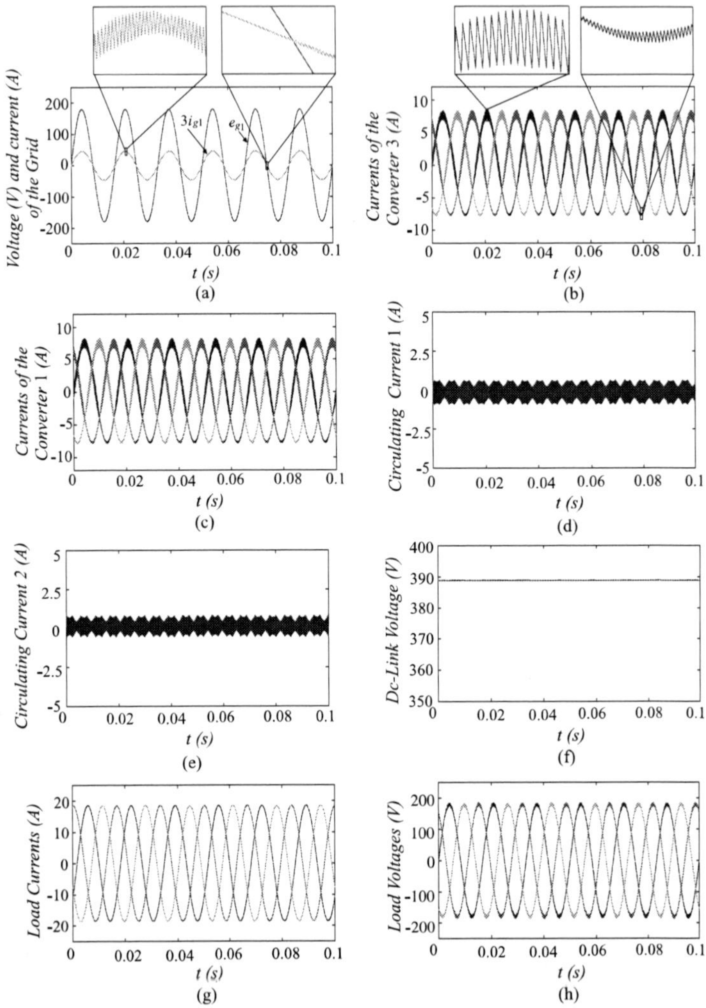

Figure 14. Simulation results for the topology T1B by using two carrier signals with $\mu = 0$. (a) Grid voltage and current for phase 1 (e_{g1} and i_{g1}). (b) Currents of the converter 1 (i_{1a}, i_{1b} and i_{1c}). (c) Currents of the converter 3 (i_{3a}, i_{3b} and i_{3c}). (d) Circulating current i_{o1}. (e) Circulating current i_{o2}. (f) Dc-link voltage (v_c). (g) Load currents (i_{l1}, i_{l2} and i_{l3}). (h) Load voltages (e_{l1}, e_{l1} and e_{l3}).

Fig. 15 illustrates the results for a single carrier signal with $\mu = 0.5$. From those

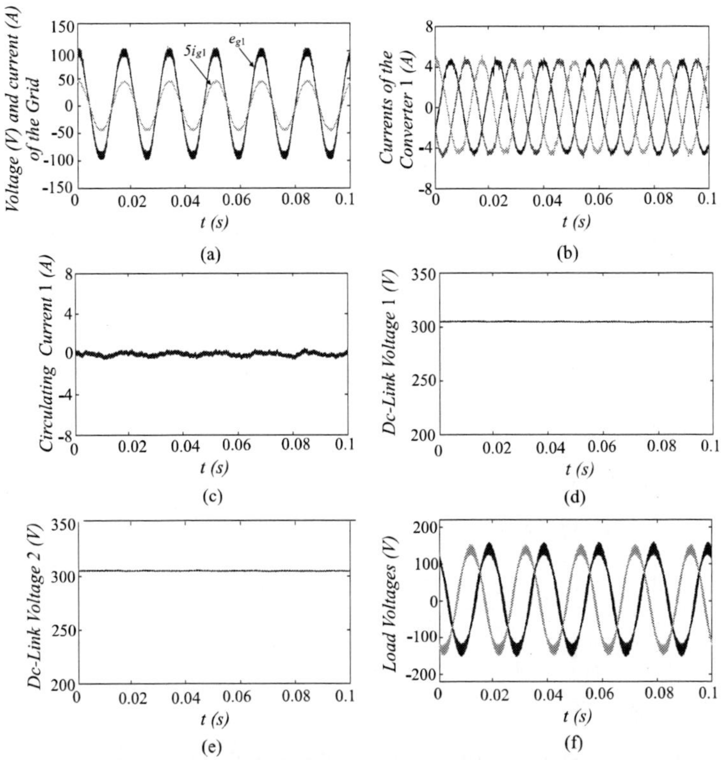

Figure 15. Experimental results for the topology T2B with a single carrier signal and $\mu = 0.5$. (a) Grid voltage and current for phase 1 (e_{g1} and i_{g1}). (b) Currents of converter 1 (i_{1a}, i_{1b} and i_{1c}). (c) Circulating current (i_{o1}). (d) Dc-link voltage 1 (v_{c1}). (e) Dc-link voltage 2 (v_{c2}). (f) Load voltages (e_{l1} and e_{l2}).

results it is observed that the current i_{g1} is a sine and it is in phase with the grid voltage for phase 1 [see Fig. 15(a)], the circulating current is close to zero [Fig. 15(c)], and the dc-link voltages along with the three-phase voltages are under control, as seen in Figs. 15(d), 15(e) and 15(f). Also, due to the connection in parallel the internal currents are half of the grid and load currents [see Fig. 15(b)].

The experimental results with two carrier signals with $\mu = 0.5$, are shown in Fig. 16. As expected it is verified a reduction of the high frequency components of the current i_{g1} and for the three-phase load voltages as compared to the single carrier signal [see Figs. 16(a) and 16(f)].

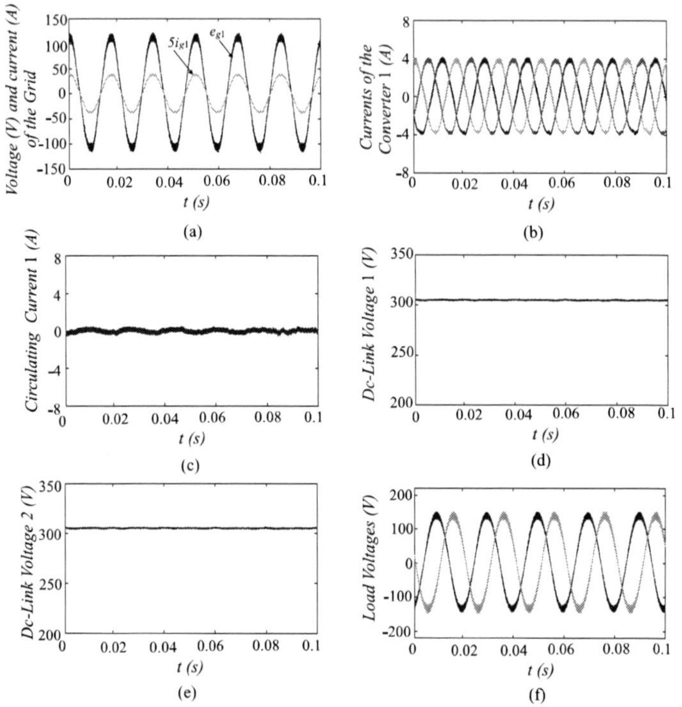

Figure 16. Experimental results of the topology T2B by using two carrier signals with $\mu = 0.5$. (a) Grid voltage and current for phase 1 (e_{g1} and i_{g1}). (b) Current of the converter 1 (i_{1a}, i_{1b} and i_{1c}). (c) Circulating current (i_{o1}). (d) Dc-link voltage 1 (v_{c1}). (e) Dc-link voltage 2 (v_{c2}). (f) Load voltages (e_{l1} and e_{l2}).

11. Comparison of the Topologies

A comparison between both proposed topologies in this chapter is conducted in this section. The best performance of the topologies T2B and T1B are met when two carrier signals are employed with either $\mu = 0$ or $\mu = 1$. Although such configurations employ twice the number of switches of the conventional back-to-back converter, the following advantages can be obtained: reduced harmonic distortion of the grid currents (57%) and load voltages (66%), reduced semiconductor losses (27%) and high frequency losses of the capacitor (23% for the topology T2B and 48% for the topology T1B), see Table 4.

Table 4. Comparison of the configurations

PWM		Conv.	T2B	T1B
Criteria		$\mu = 0.5$	$\mu = 0$ or 1	$\mu = 0$ or 1
Number of switches		12	24	24
Number of the dc-links		1	2	1
$WTHD$	i_g	1.0	0.43	0.43
	e_l	1.0	0.34	0.34
	i_{1a}	1.0	1.01	1.78
$THD(i_g)$	l_n	1.0	≥ 0.5	≥ 0.5
	f_{ch}	$10kHz$	$\geq 5kHz$	$\geq 5kHz$
$P_{c,loss}^{hf}$		1.0	0.77	0.52
P_t		1.0	0.73	0.73

12. Conclusion

The back-to-back converters presented in this chapter were characterized by the connection of two back to back units in parallel without isolation transformers. Both topologies T2B and T1B show similar performances in terms of harmonic distortions of the grid current and load voltages, as well as in terms of semiconductor losses, especially when the proposed PWM strategy is implemented with $\mu = 0$ or $\mu = 1$. In this case the proposed converters present better THD and losses than the conventional solution.

Appendix I. odq Transformation

Assuming the transformation given by:

$$\bar{w}_{123} = \bar{P}\bar{w}_{odq} \tag{72}$$

where $\bar{w}_{123} = [w_1 \;\; w_2 \;\; w_3]^T$ means vectors of voltages and currents of the system, $\bar{w}_{odq} = [w_o \;\; w_d \;\; w_q]^T$ and $\bar{P}$ is the transformation matrix defined by

$$P = \sqrt{\frac{2}{3}} \begin{bmatrix} \frac{1}{\sqrt{2}} & 1 & 0 \\ \frac{1}{\sqrt{2}} & -\frac{1}{2} & \frac{\sqrt{3}}{2} \\ \frac{1}{\sqrt{2}} & -\frac{1}{2} & -\frac{\sqrt{3}}{2} \end{bmatrix} \quad \text{and} \quad [P]^{-1} = [P]^T \tag{73}$$

Appendix II. Generalization of the Topology T2B

The topology T2B can be generalized with N back-to-back converters connected in parallel, which is constituted by Δ converters ($\Delta = 2N$), $6(N+1)$ inductive filters and N dc-link capacitors as presented in Fig. 17.

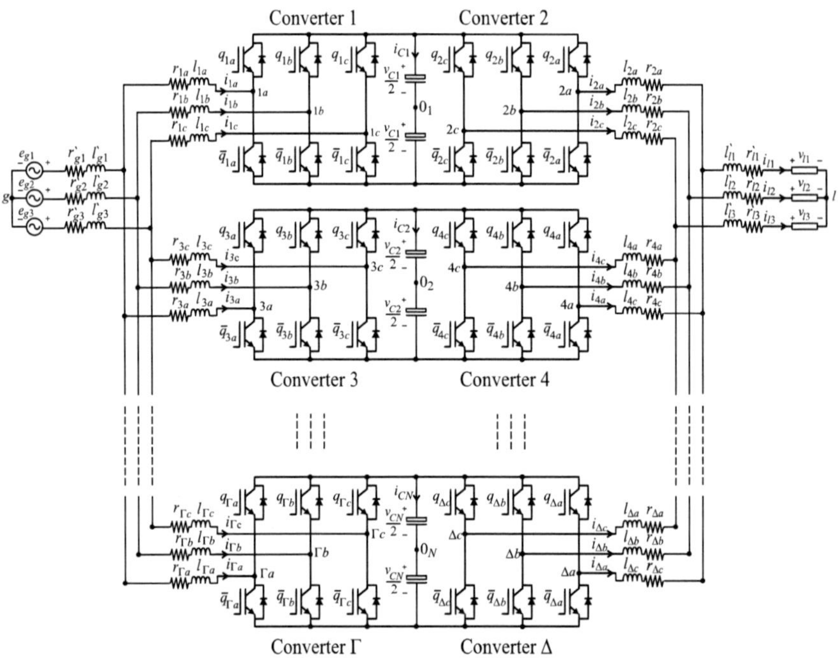

Figure 17. Generalization of the configuration T2B with N back-to-back converters connected in parallel.

12.1. Dynamic Model

The model for the generalized T2B topology is presented in the sequence. From Fig. 17 the following equations can be established:

At the grid side:

$$e_{g1} = z_g' i_{g1} + z_{\gamma a} i_{\gamma a} + v_{\gamma a0_u} - v_{g0_u} \tag{74}$$

$$e_{g2} = z_g' i_{g2} + z_{\gamma b} i_{\gamma b} + v_{\gamma b0_u} - v_{g0_u} \tag{75}$$

$$e_{g3} = z_g' i_{g3} + z_{\gamma c} i_{\gamma c} + v_{\gamma c0_u} - v_{g0_u} \tag{76}$$

with $\gamma = 1, 3, 5...\Gamma, \Gamma = 2N - 1$, $u = (\gamma + 1)/2$ and the voltage v_{g0_u} is defined by:

$$v_{g0_u} = \frac{1}{3}\sum_{j=a}^{c} v_{\gamma j0_u} \tag{77}$$

At the load side:

$$e_{l1} = -z_l' i_{l1} - z_{\delta a} i_{\delta a} + v_{\delta a0_u} - v_{l0_u} \tag{78}$$

$$e_{l2} = -z_l' i_{l2} - z_{\delta b} i_{\delta b} + v_{\delta b0_u} - v_{l0_u} \tag{79}$$

$$e_{l3} = -z_l' i_{l3} - z_{\delta b} i_{\delta c} + v_{\delta c0_u} - v_{l0_u} \tag{80}$$

with $\delta = \gamma + 1$, and the voltage v_{l0_u} is defined by:

$$v_{l0_u} = \frac{1}{3}\sum_{j=a}^{c} v_{\delta j0_u}. \tag{81}$$

In this case it is possible to write $N - 1$ meshes for the circulating current among the converters, as following:

$$v_{o\alpha} = \sum_{j=a}^{c} z_{\gamma j} i_{\gamma j} + \sum_{j=a}^{c} z_{\delta j} i_{\delta j} - \sum_{j=a}^{c} z_{\bar{\gamma} j} i_{\bar{\gamma} j} - \sum_{j=a}^{c} z_{\bar{\delta} j} i_{\bar{\delta} j} \tag{82}$$

with

$$v_{o\alpha} = -\sum_{j=a}^{c} v_{\gamma j0_u} + \sum_{j=a}^{c} v_{\delta j0_u} + \sum_{j=a}^{c} v_{\bar{\gamma} j0_{\bar{u}}} - \sum_{j=a}^{c} v_{\bar{\delta} j0_{\bar{u}}} \tag{83}$$

where $\alpha = (\gamma + 1)/2$ with $\alpha \le \Gamma - 2$, $\bar{u} = u + 1$, $\bar{\gamma} = \gamma + 2$ and $\bar{\delta} = \delta + 2$.

It is worth mentioning that the generalization for the topology T1B can be also obtained following the same procedure as done for the configuration T2B.

References

[1] M. Yilmaz and P. T. Krein. Review of the impact of vehicle-to-grid technologies on distribution systems and utility interfaces. *Power Electronics, IEEE Transactions on,* 28(12):5673-5689, Dec 2013.

[2] J. Thunes, R. Kerman, D. Schlegel, and T. Rowan. Current regulator instabilities on parallel voltage-source inverters. *IEEE Transactions on Industry Applications,* 35(1):70-77, Jan/Feb 1999.

[3] N. Tabeli, M. A. Sadrnia, and S. M. R. Rafiei. Current and voltage control of paralleled multi-module inverter systems. In *Mediterranean Conference on Control and Automation, MED*, pages 1498-1503, Thessaloniki, 24-26 June 2009. Proceedings of the 17th Mediterranean Conference.

[4] L. Asiminoaei, E. Aeloiza, J. H. Kim, P. N. E. Enjeti, and F. Blaabjerg. Parallel interleaved inverters for reactive power and harmonic compensation. In *Power Electronics Specialists Conference, PESC*, volume 2, pages 788-794, Jeju, Korea, 18-22 June 2006. Proceedings of the 37th Annual IEEE PESC.

[5] R. Ramos, D. Biel, E. Fossas, and F. Guinjoan. Interleaving quasi-sliding-mode control of parallel-connected buck-based inverters. *IEEE Transactions on Industrial Electronics*, 55(11):3865-3873, Nov. 2008.

[6] T. Kawabata and S. Higashino. Parallel operation of voltage source inverters. *IEEE Transactions on Industry Applications*, 24(2):281-287, Mar/Apr 1988.

[7] Shouchuan Chen, Xinjian Jiang, and Jianyun Chai. Analysis of harmonic and circulating current for paralleling inverters in mw-level wind energy conversion system. In *International Conference on Electrical Machines and Systems, ICEMS*, pages 2489-2492, Wuhan, 17-20 Oct. 2008. *Proceedings of the 17th ICEMS*.

[8] Ching-Tsai Pan and Yi-Hung Liao. Modeling and control of circulating currents for parallel three-phase boost rectifiers with different load sharing. *IEEE Transactions on Industrial Electronics*, 55(7):2776-2785, July 2008.

[9] Ching-Tsai Pan and Yi-Hung Liao. Modeling and coordinate control of circulating currents in parallel three-phase boost rectifiers. *IEEE Transactions on Industrial Electronics*, 54(2):825-838, April 2007.

[10] R. M. Cuzner, D. J. Nowak, A. Bendre, G. Oriti, and A. L. Julian. Mitigating circulating common-mode currents between parallel soft-switched drive systems. *IEEE Transactions on Industry Applications*, 43(5):1284-1294, sep.-oct. 2007.

[11] Di Zhang, F. Wang, R. Burgos, and D. Boroyevich. Common mode circulating current control of interleaved three-phase two-level voltage-source converters with discontinous space-vector modulation. In *Energy Conversion Congress and Exposition, ECCE*, pages 2801-2807, San Jose, CA, 20-24 Sept 2009. *Proceedings of the 1st annual IEEE ECCE.*

[12] G. Ledwich and P. Doulai. Multiple converter performance and active filtering. *IEEE Transactions on Power Electronics*, 10(3):273-279, May 1995.

[13] J. W. Dixon and B. T. Ooi. Series and parallel operation of hysteresis current-coltrolled pwm rectifiers. *IEEE Transactions on Industry Applications*, 25(4):644-651, Jul/Aug 1989.

[14] S. Ogasawara, J. Takagaki, H. Akagi, and A. Nabae. A novel control scheme of a parallel current-controlled pwm inverter. *IEEE Transactions on Industry Applications*, 28(5):1023-1030, Sep/Oct 1992.

[15] Di Zhang, F. Wang, R. Burgos, Rixin Lai, and D. Boroyevich. Impact of interleaving on ac passive components of paralleled three-phase voltagesource converters. IEEE Transactions on Industry Applications, 46(3):1042-1054, May-June 2010.

[16] K. Matsui, Y. Murai, M. Watanabe, M. Kaneko, and F. Ueda. A pulse width-modulated inverter with parallel connected transistors using current-sharing reactors. IEEE Transactions on Power Electronics, 8(2):186-191, Apr 1993.

[17] L. Asiminoaei, E. Aeloiza, P. N. Enjeti, and F. Blaabjerg. Shunt active-power-filter topology based on parallel interleaved inverters. *IEEE Trans. Industrial Electronics*, 55(3):1175-1189, March 2008.

[18] L. Asiminoaei, E. Aeloiza, J. H. Kim, P. Enjeti, F. Blaabjerg, L. T. Moran, and S. K. Sul. An interleaved active power filter with reduced size of passive components. In *Applied Power Electronics Conference and Exposition*, APEC, pages 19-23, Dallas, Texas, 19-23 March 2006. *Proceedings of the 21st Annual IEEE APEC.*

[19] Zhihong Ye, D. Boroyevich, Jae-Young Choi, and F. C. Lee. Control of circulating current in two parallel three-phase boost rectifiers. *IEEE Transactions on Power Electronics*, 17(5):609-615, Sep 2002.

[20] Zhihong Ye, D. Boroyevich, Jae-Young Choi, and F. C. Lee. Control of circulating current in parallel three-phase boost rectifiers. In *Applied Power Electronics Conference and Exposition, APEC*, volume 1, pages 506-512, New Orleans, LA, 06-10 Feb. 2000. *Proceedings of the 15th Annual IEEE APEC.*

[21] Kun Xing, F.C. Lee, D. Borojevic, Zhihong Ye, and S. Mazumder. Interleaved pwm with discontinuous space-vector modulation. *IEEE Transactions on Power Electronics*, 14(5):906-917, Sep 1999.

[22] Yang Chen and K.M. Smedley. Parallel operation of one-cycle controlled three-phase pfc rectifiers. *IEEE Transactions on Industrial Electronics*, 54(6):3217-3224, Dec. 2007.

[23] R. Giral, L. Martinez-Salamero, and S. Singer. Interleaved converters operation based on cmc. *IEEE Transactions on Power Electronics*, 14(4):643-652, Jul 1999.

[24] C. B. Jacobina, A. M. N Lima, E. R. C. da Silva, R. N. C. Alves, and P. F. Seixas. Digital scalar pulse-width modulation: A simple approach to introduce non-sinusoidal modulating waveforms. *IEEE Transactions on Power Electronics*, 16(3):351-359, May 2001.

[25] V. Blasko. Analysis of a hybrid pwm based on modified space-vector and triangle-comparison methods. *IEEE Transactions on Industry Applications*, 33(3):756-764, May/Jun 1997.

[26] Keliang Zhou and Danwei Wang. Relationship between space-vector modulation and three-phase carrier-based pwm: A comprehensive analysis [three-phase inverters]. *IEEE Trans. Ind. Electron.*, 49(1):186-196, Feb 2002.

[27] M.C. Cavalcanti, E.R. da Silva, D. Boroyevich, W. Dong, and C. B. Jacobina. A feasible loss model for igbt in soft-switching inverters. In *Power Electronics Specialist Conference, PESC IEEE 34th Annual*, volume 3, pages 1845-1850, 15-19 June 2003. *Proceedings of the 34th Annual IEEE PESC.*

[28] F. D. Kieferndorf, M. Forster, and T. A. Lipo. Reduction of dc-bus capacitor ripple current with pam/pwm converter. *IEEE Trans. Industry Applications*, 40(2):607-614, March-April 2004.

[29] K. Harada, A. Katsuki, and M. Fujiwara. Use of esr for deterioration diagnosis of electrolytic capacitor. *IEEE Transactions on Power Electronics*, 8(4):355-361, Oct 1993.

[30] A. M. R. Amaral and A. Cardoso. An economic offline technique for estimating the equivalent circuit of aluminum electrolytic capacitors. *IEEE Transactions on Instrumentation and Measurement*, 57(12):2697-2710, Dec. 2008.

[31] W. Ming Chen, Hsu-Chin Wu, Ming-Wei Chou, and Kung-Yen Lee. Online failure prediction of the electrolytic capacitor for lc filter of switching-mode power converters. *IEEE Transactions on Industrial Electronics*, 55(1):400-406, Jan. 2008.

[32] M.L. Gasperi. A method for predicting the expected life of bus capacitors. In *Industry Applications Conference*, IAS, volume 2, pages 1042-1047, New Orleans, LA, 05-06 Oct 1997. *Proceedings of the 32nd IAS Annual Meeting.*

[33] P. Venet, F. Perisse, M. H. El-Husseini, and G. Rojat. Realization of a smart electrolytic capacitor circuit. *Industry Applications, IEEE Magazine*, pages 16-205, Jan, Feb 2002.

[34] Evox-Rifa. Electrolytic capacitors application guide, 2001. Disponível em ¡http: // www.evox-rifa.com¿ acessado em maio de 2010.

[35] J. W. Kolar and S. D. Round. Analytical calculation of the rms current stress on the dc-link capacitor of voltage-pwm converter systems. *Electric Power Applications, IEE Proceedings* -, 153(4):535-543, July 2006.

INDEX

#

A

B

C

D

E

H

I

J

K

L

M

N

Q

R

S

T

U

V

W

X

Y

Z